Outlooks

Readings for Environmental Literacy

Second Edition

Edited by
Michael L. McKinney
University of Tennessee, Knoxville

JONES AND BARTLETT PUBLISHERS
Sudbury, Massachusetts
BOSTON   TORONTO   LONDON   SINGAPORE

World Headquarters

Jones and Bartlett Publishers
40 Tall Pine Drive
Sudbury, MA 01776
978-443-5000
info@jbpub.com
www.jbpub.com

Jones and Bartlett Publishers Canada
2406 Nikanna Road
Mississauga, ON L5C 2W6
CANADA

Jones and Bartlett Publishers
International
Barb House, Barb Mews
London W6 7PA
UK

ISBN 0-7637-3280-X

Production Credits
Executive Editor, Science: Stephen L. Weaver
Managing Editor, Science: Dean W. DeChambeau
Senior Production Editor: Louis C. Bruno, Jr.
Marketing Manager: Matthew Bennett
Marketing Associate: Matthew Payne
Text Design: Anne Spencer
Composition: International Typesetting and Composition
Cover Design: Kristin Ohlin
Printing and Binding: Courier Stoughton
Cover Printing: Courier Stoughton

Acknowledgments
The editors and Jones and Bartlett Publishers thank the many freelance authors, magazines, and journals for allowing us to reprint their articles.

Cover image: AbleStock

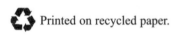 Printed on recycled paper.

CONTENTS

SECTION ONE: THE ENVIRONMENT AND HUMANS   **1**

1. **The Challenges We Face 3**
   Jeffrey Kluger and Andrea Dorfman (*Time*, August 26, 2002)

2. **Perceiving the Population Bomb 7**
   Andrew R. B. Ferguson (*World Watch*, July/August 2001)

3. **Rich vs. Poor 10**
   Nicole Itano (*E: the Environmental Magazine*, November/December 2002)

4. **Environmental Refugees 12**
   Mark Townsend (*The Ecologist*, July/August 2002)

SECTION TWO: ENVIRONMENT OF LIFE ON EARTH   **17**

5. **The Most Important Fish in the Sea 19**
   H. Bruce Franklin (*Discover*, September 2001)

6. **Trout Are Wildlife, Too 24**
   Ted Williams (*Audubon*, December 2002)

7. **Hostile Beauty 29**
   Geoffrey O'Gara (*National Wildlife*, August/September 2002)

8. **Wilding America 32**
   Elizabeth Royte (*Discover*, September 2002)

SECTION THREE: RESOURCE USE AND MANAGEMENT   **37**

9. **The Winds of Change 39**
   Margot Roosevelt (*Time*, August 26, 2002)

10. **Scientists Say a Quest for Clean Energy Must Begin Now 42**
    Andrew C. Revkin (*New York Times*, November 1, 2002)

11. **Link Seen Between Water Scarcity and Poverty 44**
    Sanjay Suri (*Global Information Network*, December 12, 2002)

12. **Atlanta's Growing Thirst Creates Water War 47**
    Douglas Jehl (*New York Times*, May 27, 2002)

13. **North America Losing Biodiversity, Say Experts 50**
    Danielle Knight (*Global Information Network*, January 7, 2002)

14. **Buzz Cut 52**
    Paul Rauber (*Sierra*, April 2003)

15. **Feeding the World 55**
    Luther Tweeten, Carl Zulauf (*The Futurist*, September/October 2002)

SECTION FOUR: DEALING WITH ENVIRONMENTAL DEGRADATION   **61**

16. **Ill Winds: The Chemical Plant Next Door 63**
    Becky Bradway (*E: The Environmental Magazine*, September/October 2002)

17. **Facing Up to a Dirty Secret 68**
    Erling Hoh (*Far Eastern Economic Review*, December 12, 2002)

18. **Attacking an Arsenic Plague 71**
Helen Epstein (*Popular Science*, November 2002)

19. **Long-Term Data Show Lingering Effects from Acid Rain 73**
Kevin Krajick (*Science*, April 13, 2001)

20. **News on the Environment Isn't Always Bad 76**
Mark Sappenfield (*Christian Science Monitor*, October 4, 2002)

21. **The Weather Turns Wild 78**
Nancy Shute, Thomas Hayden, Charles W. Petit, Rachel K. Sobel, Kevin Whitelaw, David Whitman (*U.S. News & World Report*, February 5, 2001)

22. **Climate Policy Needs a New Approach 82**
David Applegate (*Geotimes,* May 2001)

23. **Bioreactors and EPA Proposal to Deregulate Landfills 85**
Bill Sheehan, Jim McNelly (*BioCycle*, January 2003)

24. **Managing the Environmental Legacy of U.S. Nuclear-Weapons Production 88**
Kevin D. Crowley, John F. Ahearne (*American Scientist*, November/December 2002)

25. **Silent Spring: A Sequel? 96**
Les Line (*National Wildlife*, December 2002/January 2003)

SECTION FIVE: ENVIRONMENTAL ISSUES: ASPECTS AND SOLUTIONS   101

26. **Too Green for Their Own Good? 103**
Andrew Goldstein (*Time*, August 26, 2002)

27. **Seeing Green: Knowing and Saving the Environment on Film 106**
Luis A. Vivanco (*American Anthropologist*, December 2002)

28. **A Forest Path Out of Poverty 117**
Arie Farnam (*Christian Science Monitor*, August 9, 2002)

29. **Growers and Greens Unite 119**
Gerald Haslam (*Sierra,* January/February 2003)

30. **Needed: A National Center for Biological Invasions 123**
Don C. Schmitz, Daniel Simberloff (*Issues in Science and Technology*, Summer 2001)

31. **Privatizing Water 128**
Curtis Runyan (*World Watch*, January/February 2003)

32. **Groups Sue Government Agency Over Global Warming 131**
Jim Lobe (*Global Information Network*, December 5, 2002)

33. **GM and Ford Pressed to Cut Greenhouse Gases 133**
Jim Lobe (*Global Information Network*, December 12, 2002)

34. **Tricks of Free Trade 135**
Mark Weisbrot (*Sierra,* September/October 2001)

35. **Lots of It About–Corporate Social Responsibility 139**
The Economist, December 14, 2002

36. **Economic Growth and the Environment: Alternatives to the Limits Paradigm 142**
Carlos Davidson (*BioScience*, May, 2000)

Answers 151

The Environment and Humans

World leaders converged for the World Summit on Sustainable Development in Johannesburg, South Africa, during the summer of 2002. In the 10 years since Rio, the environmental problems are mostly the same. The world population is still rising and is expected to reach 11 billion within the century. A third of the world is in danger of starving, while present forms of agriculture are very unsustainable. At least 1.1 billion people lack clean drinking water, a figure that is expected to rise to two-thirds of the world's population in 50 years. Climate change and biodiversity are major concerns with 11,000 species vulnerable to extinction. As the state of the environment is not bettering much, world leaders are looking toward a more holistic approach.

The Challenges We Face

Jeffrey Kluger, Andrea Dorfman

Time, August 26, 2002 (Special Report: Green Century)

FOR STARTERS, LET'S BE CLEAR about what we mean by "saving the earth." The globe doesn't need to be saved by us, and we couldn't kill it if we tried. What we do need to save—and what we have done a fair job of bollixing up so far—is the earth as we like it, with its climate, air, water and biomass all in that destructible balance that best supports life as we have come to know it. Muck that up, and the planet will simply shake us off, as it's shaken off countless species before us. In the end, then, it's us we're trying to save—and while the job is doable, it won't be easy.

The 1992 Earth Summit in Rio de Janeiro was the last time world leaders assembled to look at how to heal the ailing environment. Now, 10 years later, Presidents and Prime Ministers are convening at the World Summit on Sustainable Development in Johannesburg next week to reassess the planet's condition and talk about where to go from here. In many ways, things haven't changed: the air is just as grimy in many places, the oceans just as stressed, and most treaties designed to do something about it lie in incomplete states of ratification or implementation. Yet we're oddly smarter than we were in Rio. If years of environmental false starts have taught us anything, it's that it's time to quit seeing the job of cleaning up the world as a zero-sum game between industrial progress on the one hand and a healthy planet on the other. The fact is, it's development—well-planned, well-executed sustainable development—that may be what saves our bacon before it's too late.

As the summiteers gather in Johannesburg, TIME is looking ahead to what the unfolding century—a green century—could be like. In this special report, we will examine several avenues to a healthier future, including green industry, green architecture, green energy, green transportation and even a greener approach to wilderness preservation. All of them have been explored before, but never so urgently as now. What gives such endeavors their new credibility is the hope and notion of sustainable development, a concept that can be hard to implement but wonderfully simple to understand.

With 6.1 billion people relying on the resources of the same small planet, we're coming to realize that we're drawing from a finite account. The amount of crops, animals and other biomatter we extract from the earth each year exceeds what the planet can replace by an estimated 20%, meaning it takes 14.4 months to replenish what we use in 12—deficit spending of the worst kind. Sustainable development works to reverse that, to expand the resource base and adjust how we use it so we're living off biological interest without ever touching principal. "The old environmental movement had a reputation of elitism," says Mark Malloch Brown, administrator of the United Nations Development Program (UNDP). "The key now is to put people first and the environment second, but also to remember that when you exhaust resources, you destroy people." With that in mind, the summiteers will wrestle with a host of difficult issues that affect both people and the environment. Among them:

Population and Health: While the number of people on earth is still rising rapidly, especially in the developing countries of Asia, the good news is that the growth rate is slowing. World population increased 48% from 1975 to 2000, compared with 64% from 1950 to 1975. As this gradual deceleration continues, the population is expected to level off eventually, perhaps at 11 billion sometime in the last half of this century.

Economic-development and family-planning programs have helped slow the tide of people, but in some places, population growth is moderating for all the wrong reasons. In the poorest parts of the world, most notably Africa, infectious diseases such as AIDS, malaria, cholera and tuberculosis are having a Malthusian effect. Rural-land degradation is pushing people into cities, where crowded, polluted living conditions create the perfect breeding grounds for sickness. Worldwide, at least 68 million are expected to die of AIDS by 2020, including 55 million in sub-Saharan Africa. While any factor that eases population pressures may help the environment, the situation would be far less tragic if rich nations did more to help the developing world reduce birth rates and slow the spread of disease.

Efforts to provide greater access to family planning and health care have proved effective. Though women in the poorest countries still have the most children, their collective fertility rate is 50% lower than it was in 1969 and is expected to decline more by 2050. Other programs targeted at women include basic education and job training. Educated mothers not only have a stepladder out of poverty, but they also choose to have fewer babies.

Rapid development will require good health care for the young since there are more than 1 billion people ages 15 to 24. Getting programs in place to keep this youth bubble healthy could make it the most productive generation ever conceived. Says Thoraya Obaid, executive director of the U.N. Population Fund: "It's a window of opportunity to build the economy and prepare for the future."

Food: Though it's not always easy to see it from the well-fed West, up to a third of the world is in danger of starving. Two billion people lack reliable access to safe, nutritious food, and 800 million of them—including 300 million children— are chronically malnourished.

Agricultural policies now in place define the very idea of unsustainable development. Just 15 cash crops such as corn, wheat and rice provide 90% of the world's food, but planting and replanting the same crops strips fields of nutrients and makes them more vulnerable to pests. Slash-and-burn planting techniques and overreliance on pesticides further degrade the soil.

Solving the problem is difficult, mostly because of the ferocious debate over how to do it. Biotech partisans say the answer lies in genetically modified crops—foods engineered for vitamins, yield and robust growth. Environmentalists worry that fooling about with genes is a recipe for Frankensteinian disaster. There is no reason, however, that both camps can't make a contribution.

Better crop rotation and irrigation can help protect fields from exhaustion and erosion. Old-fashioned cross-breeding can yield plant strains that are heartier and more pest-resistant. But in a world that needs action fast, genetic engineering must still have a role—provided it produces suitable crops. Increasingly, those crops are being created not just by giant biotech firms but also by home-grown groups that know best what local consumers need.

The National Agricultural Research Organization of Uganda has developed corn varieties that are more resistant to disease and thrive in soil that is poor in nitrogen. Agronomists in Kenya are developing a sweet potato that wards off viruses. Also in the works are drought-tolerant, disease-defeating and vitamin-fortified forms of such crops as sorghum and cassava—hardly staples in the West, but essentials elsewhere in the world. The key, explains economist Jeffrey Sachs, head of Columbia University's Earth Institute, is not to dictate food policy from the West but to help the developing world build its own biotech infrastructure so it can produce the things it needs the most. "We can't presume that our technologies will bail out poor people in Malawi," he says. "They need their own improved varieties of sorghum and millet, not our genetically improved varieties of wheat and soybeans."

Water: For a world that is 70% water, things are drying up fast. Only 2.5% of water is fresh, and only a fraction of that is accessible. Meanwhile, each of us requires about 50 quarts per day for drinking, bathing, cooking and other basic needs. At present, 1.1 billion people lack access to clean drinking water and more than 2.4 billion lack adequate sanitation. "Unless we take swift and decisive action," says U.N. Secretary-General Kofi Annan, "by 2025, two-thirds of the world's population may be living in countries that face serious water shortages."

The answer is to get smart about how we use water. Agriculture accounts for about two-thirds of the fresh water consumed. A report prepared for the summit thus endorses the "more crop per drop" approach, which calls for more efficient irrigation techniques, planting of drought- and salt-tolerant crop varieties that require less water and better monitoring of growing conditions, such as soil humidity levels. Improving water-delivery systems would also help, reducing the amount that is lost en route to the people who use it.

One program winning quick support is dubbed WASH—for Water, Sanitation and Hygiene for All—a global effort that aims to provide water services and hygiene training to everyone who lacks them by 2015. Already, the U.N., 28 governments and many nongovernmental organizations (ngos) have signed on.

Energy and Climate: In the U.S., people think of rural electrification as a long-ago legacy of the New Deal. In many parts of the world, it hasn't even happened yet. About 2.5 billion people have no access to modern energy services, and the power demands of developing economies are expected to grow 2.5% per year. But if those demands are met by burning fossil fuels such as oil, coal and gas, more and more carbon dioxide and other greenhouse gases will hit the atmosphere. That, scientists tell us, will promote global warming, which could lead to rising seas, fiercer storms, severe droughts and other climatic disruptions.

Of more immediate concern is the heavy air pollution caused in many places by combustion of wood and fossil fuels. A new U.N. Environment Program report warns of the effects of a haze across all southern Asia. Dubbed the "Asian brown cloud" and estimated to be 2 miles thick, it may be responsible for hundreds of thousands of deaths a year from respiratory diseases.

The better way to meet the world's energy needs is to develop cheaper, cleaner sources. Pre-Johannesburg proposals call for eliminating taxation and pricing systems that encourage oil use and replacing them with policies that provide incentives for alternative energy. In India there has been a boom in wind power because the government has made it easier for entrepreneurs to get their hands on the necessary technology and has then required the national power grid to purchase the juice that wind systems produce.

Other technologies can work their own little miracles. Micro-hydroelectric plants are already operating in numerous nations, including Kenya, Sri Lanka and Nepal. The systems divert water from streams and rivers and use it to run turbines without complex dams or catchment areas. Each plant can produce as much as 200 kilowatts—enough to electrify 200 to 500 homes and businesses—and lasts 20 years. One plant in Kenya was built by 200 villagers, all of whom own shares in the cooperative that sells the power.

The Global Village Energy Partnership, which involves the World Bank, the UNDP and various donors, wants to provide energy to 300 million people, as well as schools, hospitals and clinics in 50,000 communities worldwide over 10 years. The key will be to match the right energy source to the right users. For example, solar panels that convert sunlight into electricity might be cost-effective in remote areas, while extending the power grid might be better in Third World cities.

Biodiversity: More than 11,000 species of animals and plants are known to be threatened with extinction, about a third of all coral reefs are expected to vanish in the next 30 years and about 36 million acres of forest are being razed annually. In his new book, The Future of Life, Harvard biologist Edward O. Wilson writes of his worry that unless we change our ways half of all species could disappear by the end of this century.

The damage being done is more than aesthetic. Many vanishing species provide humans with both food and medicine. What's more, once you start tearing out swaths of ecosystem, you upset the existing balance in ways that harm even areas you didn't intend to touch. Environmentalists have said this for decades, and now that many of them have tempered ecological absolutism with developmental realism, more people are listening.

The Equator Initiative, a public-private group, is publicizing examples of sustainable development in the equatorial belt. Among the projects already cited are one to help restore marine fisheries in Fiji and another that promotes beekeeping as a source of supplementary income in rural Kenya. The Global Conservation Trust hopes to raise $260 million to help conserve genetic material from plants for use by local agricultural programs. "When you approach sustainable development from an environmental view, the problems are global," says the U.N.'s Malloch Brown. "But from a development view, the front line is local, local, local."

If that's the message environmental groups and industry want to get out, they appear to be doing a good job of it. Increasingly, local folks act whether world political bodies do or not. California Governor Gray Davis signed a law last month requiring automakers to cut their cars' carbon emissions by 2009. Many countries are similarly proactive. Chile is encouraging sustainable use of water and electricity; Japan is dangling financial incentives before consumers who buy environmentally sound cars; and tiny Mauritius is promoting solar cells and discouraging use of plastics and other disposables.

Business is getting right with the environment too. The Center for Environmental Leadership in Business, based in Washington, is working with auto and oil giants including Ford, Chevron, Texaco and Shell to draft guidelines for incorporating biodiversity conservation into oil and gas exploration. And the center has helped Starbucks develop purchasing guidelines that reward coffee growers whose methods have the least impact on the environment. Says Nitin Desai, secretary-general of the Johannesburg summit: "We're hoping that partnerships—involving governments, corporations, philanthropies and NGOs—will increase the credibility of the commitment to sustainable development."

Will that happen? In 1992 the big, global measures of the Rio summit seemed like the answer to what ails the world. In 2002 that illness is—in many respects—worse. But if Rio's goal was to stamp out the disease of environmental degradation, Johannesburg's appears to be subtler—and perhaps better: treating the patient a bit at a time, until the planet as a whole at last gets well. ❖

Questions

1. Why did world leaders assemble in Rio de Janeiro in 1992?
2. How many people currently lack access to clean drinking water?
3. How many species are known to be threatened with extinction?

Answers are at the back of the book.

In 1968, few scholars were ready to accept Paul Ehrlich's warnings of the effects of a population explosion or that a population problem was even occurring. Over the past 32 years, however, demographers have come to understand that his warning was very real, and the effects are surfacing. "Collateral damage" from the "population bomb" includes a degraded environment and dwindling resources. In 1993, a biophysical scientist, Vaclav Smil, provided the graphic evidence necessary to show exactly when the population bomb went off by comparing the drop in infant mortality to a rise in energy consumption.

2

Perceiving the Population Bomb

Andrew R.B. Ferguson

World Watch, Washington, July/August 2001

THIRTY-TWO YEARS AGO, in 1968, Paul Ehrlich sounded a wake-up call to the world with his book The Population Bomb. Now that we can see the Bomb in historical perspective, even establishing when it went off, let us set ourselves the task of perceiving-with rigorous objectivity-the explosion of the Bomb and the collateral damage it has caused.

It was only in the 1990s that many demographers came to realize that Ehrlich's message was essentially correct. Writing in 1991, Clive Pointing started the preface of his great book, A Green History of the World, with these words: "As some people climb mountains because they are there, others find themselves writing books because they are not there." His book was indeed ground-breaking; before it was published, few people can have had an adequate grasp of humanity's impact on the environment. Two years later, in 1993 the biophysical scientist, Vaclav Smil, wrote: "We have at least started to realize the enormity of environmental transformation which is imperiling the survival of modern civilization." Smil identified a spectrum of critical changes taking place in the Earth's condition, and noted that "these changes can be ordered into three broad categories: declining availability of critical natural resources and services; changing composition of the atmosphere; and the loss of biodiversity." [Italics added.] It may be impossible to rank those three in importance. However, we can use the "changing composition of the atmosphere" to establish when the Bomb went off, so let us look at that first.

In 1990, the world was emitting, from the burning of fossil fuels and cement production, about 4.2 tons of carbon dioxide per person, indicating an energy use of about 64 GJ (gigajoules) per person per year. Sixty-four GJ/capita, as well as being the average use of energy in 1990, is a reasonable lower limit for average energy use. Vaclav Smil provides the evidence. He shows us graphically that infant mortality drops precipitously as per capita energy use rises to 50 GJ per year. Thereafter, further improvement becomes more gradual. Of course there is no clear changeover point, but 64 gigajoules a year per person seems a judicious minimum. In any case, should you want to argue for a lower average use of energy than 64 GJ/year, then you will need to argue for accepting an infant mortality higher than 15 per 1000 births. Countries like Greece, Japan, and the United Kingdom, which have better infant mortality rates than that, also have energy use above 64 GJ/capita/year. Of course a correlation does not prove a causative relationship, but Smil also shows that good access to post-secondary education (20 percent of eligible population) generally involves per capita energy use above 70 GJ/year. It does seem inherently likely, moreover, that a certain minimum energy availability is necessary to achieve these benchmarks.

Assuming there is some minimum per capita emission figure, then, just how much carbon dioxide can the Earth tolerate? In order to stabilize the amount of carbon dioxide in the atmosphere, the Intergovernmental Panel on Climate Change has estimated that carbon dioxide emissions from fossil fuel would have to be reduced by at least 60 percent from their 1990 level of 22.3 billion tons, which would yield an upper limit of 8.9 billion tons. Thus the maximum population which the Earth can accommodate, while allowing the aforesaid carbon dioxide emission of 4.2 tons per person per year, is 8.9/4.2=2.1 billion people. World population reached 2.1 billion in 1940, so 1940 is the time that the Bomb went off. Collateral damage has been steadily accruing ever since, in the form of atmospheric disturbance with its corollaries, such as melting glaciers, droughts, floods and increasing frequency of the El Nino phenomenon.

By the 1990s, the tide of humanity had swept across the globe appropriating to its own use the most productive ecosystems (see map, inside back cover). By then, Smil tells us, there was only about one-third of the continental surface left over for the use of other life forms. Of this remnant, he says, "most of this area is in highly stressed low-productivity eco-systems. About 40 percent of the total are in circumpolar tundra communities, and another 20 percent are in subtropical deserts and semi-deserts, while only about 6 percent of the remaining wilderness during the mid-1980s was in tropical rain forests, and less than 2 percent in temperate (rain and broadleaf) forests." Thus we have not been generous to other species and, as Smil observes, "There is general agreement among the students of conservation biology that most existing national parks and reserves are far too small to guarantee long-term survival of especially large species."

Of course not all the damage occurred since the Bomb went off, but since the world has acquired two thirds of its human population since that time, it is likely that most of it did. Moreover, since the rate of damage has doubtless been accelerating, there is plausibility in the thesis that the human race is likely to cause an extirpation of species comparable to that of the previous five mass extinction events. Such, then, is our perception of collateral damage with respect to Smil's category "loss of biodiversity."

The remaining category mentioned by Smil is "declining availability of critical natural resources and services."

For assessing that, let us turn first to Colin Campbell, who estimates that the world, currently using about four cubic kilometers of crude oil a year, has used about half of its conventional oil and close to half of its natural gas. Moreover, we learn from Richard Duncan that both oil supply per capita and energy supply per capita peaked in 1979. But for most of the details of collateral damage in this category, it would be best to turn to a 1999 paper by David Pimentel et al. Incidentally, this paper also proposes the need to aim for a world population of 2 billion, although for different reasons from carbon dioxide emissions. From it, we can add the following items to our list (all thoroughly referenced in the original): (a) "3 billion humans malnourished worldwide"; (b) "40,000 children die each day due to malnutrition and other diseases"; (c) "Globally, the annual loss of land to urbanization and highway ranges from 10 to 35 million hectares per year, with half of this lost land coming from cropland"; (d) "Worldwide, more than 10 million hectares of productive arable land are severely degraded and abandoned each year" (about 7 percent of the total in a decade); (e) "Water demands already far exceed supplies in nearly 80 nations of the world"; (f) Since 1960, "nearly one-third of the world's arable land has been lost due to urbanization, highways, soil erosion, salinization, and water logging of the soil"; (g) "grain production per capita started declining in 1984 and continues to decline"; (h) "irrigation per capita started to declining in 1978 and continues to fall"; (i) "food production per capita started declining in 1980 and continues to fall"; (j) "fertilizer supplies essential for food production started declining in 1989 and continue to do so." That abbreviated shortlist, which omits loss of soil-serious but difficult to measure-suffices to indicate the importance of the last of Smil's three categories, "declining availability of critical natural resources and services." However, perhaps we should also recall the net loss of forest, equal to an area of at least 12 million hectares (130 miles by 130 miles) per year. We can, for the same reasons as previously, deduce that most of the damage has happened since the Bomb went off in 1940.

So, that concludes our attempt to perceive the Population Bomb, Let us now leave the hard "facts," and muse on more speculative matters. Because the Bomb exploded some 30 years before Ehrlich's book, we might be inclined to blame him and his fellow ecologists for not sounding the wake-up call earlier. However were they to have done so, the message would have fallen on deaf ears. Even some of the most intelligent people could not see that there was a population problem, and some were unwilling even to concede that a problem loomed over the horizon. And today there are few ready to listen to those who have, for several decades, been toiling to awaken the world to our perilous situation.

The efforts of some people have been superlative. In addition to the work of Paul and Anne Ehrlich, Clive Ponting, Colin Campbell, and David and Marcia Pimentel, we have had important contributions from Virginia Abernethy, Albert Bartlett, Lester Brown, Sandra Bukkens, Gretchen Daily, Richard Duncan, Robert Engleman, Garrett Hardin, Mario Giampietro, Norman Myers, Jack Parsons, Peter Tod, Mary White, and Walter Youngquist. Although it is hard to say whether the course of history has been influenced by the many fine lectures these people have given (1,200 from Al Bartlett alone), and the equally superb papers and books they have written, the fact that homo sapiens has produced such farsighted individuals-striving to abate our deep-dyed procreational proclivity-gives some cause for admiration of the human race, whatever one may think of the chances of the species surviving its incompetence in matters of fertility. ❖

Source: Worldwatch Institute, *World Watch*, Vol 14, No. 4, copyright 2001, www.worldwatch.org

Questions

1. How much carbon dioxide per person was the world emitting in 1990?
2. What is the maximum population that the Earth can accommodate, while allowing carbon dioxide emissions of 4.2 tons per person per year?
3. Both oil supply per capita and energy supply per capita peaked in what year?

Answers are at the back of the book.

3

As the World Summit on Sustainable Development of 2002 drew to a close, critiques of the summit commenced. Critics argued that reproductive policy and population woes were not mentioned. Additionally, there was "little talk" concerning reductions in greenhouse emissions to alleviate global warming. They further pointed out that topics such as recycling and renewable energy were not given much discussion time. However, the biggest complaint was that there was minimal effort put forth toward resolving the disparity factor between nations.

Rich vs. Poor

Nicole Itano

E: *The Environmental Magazine*, November/December 2002

SOUTH AFRICA IS A COUNTRY OF FABULOUS WEALTH and grinding poverty, but few delegates to the World Summit on Sustainable Development held in Johannesburg, South Africa August 26 through September 4 saw much more of the city than the malls and banking halls of one of the continent's richest neighborhoods. They also ate. According to the British Sun newspaper, the 60,000 delegates dined on delicacies, including 4,400 pounds of fillet steak, 450 pounds of salmon, more than 1,000 pounds of lobster and shellfish, 1,000 pounds of bacon and sausages, buckets of caviar and piles of pate de foie gras.

The summit, the largest of its kind ever under United Nations auspices, was supposed to build on the progress made since the Rio Earth Summit 10 years ago. But activists say the government officials from 190 nations who hammered out the Johannesburg Plan of Action behind a wall of police and barricades forgot the connection between this summit and the last. As Tom Turner of Earthjustice put it, "People will be sifting through the ashes for some time, but a milestone this was not. If the environment is to survive, it will be despite this conference, rather than because of it."

"We saw a significant backtracking at this summit in terms of real principles," says Michael Dorsey, a Sierra Club board member. "The most important of those principles was that we have an obligation to work together to solve the world's problems." For many environmentalists, the best thing out of Jo'burg was the agreement to halve the number of people without clean drinking water or sanitation by 2015.

Nowhere in the conference statement is population mentioned, largely activists say because of the Bush Administration's hostility to family planning and abortion. But the world has grown from 1.6 billion people in 1900 to more than six billion today. Most of that growth has taken place in the developing world, and analysts say that if all those people suddenly began consuming resources at a western level, the environmental effects would be devastating.

"If we are going to be serious about the summit goals and the goal of human dignity, it's essential that we start talking about issues of reproductive policy and population," said former U.S. Senator Timothy Wirth (DCO), who heads Ted Turner's United Nations Foundation.

Ironically, the only organization talking about population and consumption rates in Johannesburg was the World Bank. Estimating that the world's economy will triple to $140 trillion by 2050, the bank warned, "A $140 trillion world simply cannot rely on the current production and consumption patterns."

There was also not much talk in Johannesburg about global warming reduction, recycling or any of the other resource issues discussed at Rio. Energy was at least on the agenda, but efforts to set a definite timetable for conversion to renewables were thwarted. Only a goal of ensuring access to energy for 35 percent of the African population by 2022 remained.

Without tackling the growing consumption of the developed world and the growing numbers in the developing, however, many green groups say the summit did little to protect the environment. Some dubbed it "Rio Minus 10." Environmental groups are particularly upset about the short shrift given the Kyoto Treaty, which was abandoned by President George W. Bush earlier this year. "Kyoto gets to the heart of consumption issues, because it encourages the development of dean societies and recognizes that the developed world is in the best position to do something to reverse the effects of climate change," says Don Henry, executive director of the Australian Conservation Fund. "We need global action on climate change now." A resolution calling for Kyoto ratification was finally issued, but it's unlikely to have much effect on a recalcitrant U.S.-the most significant holdout.

Two miles from the convention center lies one of Johannesburg's poorest slums, Alexandra, where tin shacks line the banks of the polluted Jukski river and children line up for a drink at open standpipes. Alexandra resident Zakhele Lengoati says he's glad the summit took place in his city because it focused attention on the plight of his neighborhood. "For them to see Alexandra the way it is will help change it in the future," he says. Lengoati's hope that this summit will help raise the poor to the standard of the rich was a prominent theme. As the sign of one South African protester noted, the poor want development, sustainable or not.

As many as 10,000 demonstrators, including many Alexandra residents and foreign activists, took to the streets outside the conference, waving placards that read "Land, Food, Jobs" and "Water: A Human Right." Said one activist, "Most of us have given up on the summit process as another greenwash."

Host nation South Africa has put the emphasis on new investment; meanwhile, developing countries were hoping to get commitments from industrial nations to help fund dean water, sanitation and health programs.

Opening the summit, South African President Thabo Mbeki called for "shared prosperity," and said that "a global human society based on poverty for many and prosperity for a few, characterized by islands of wealth and surrounded by a sea of poverty, is unsustainable." Mbeki is probably the world's most prominent advocate of this point of view and has spent recent months trying to get support for his New Economic Plan for African Development, which promotes international investment.

Mbeki has found support in many developing countries, which feel left out of the globalization process. Much of the debate in Johannesburg sounded more like trade talk than environment talk, with issues like the reduction of First World subsidies and tariffs on the top of the agenda. Developing countries say they would be less reliant on international aid if the European Union and United States would level the trade playing field.

The focus on the free market worried many environmentalists, who fear that free trade principles will supersede national environmental regulations. While they agree that trade barriers against poor nations should be dropped, they say free trade is not the answer to the world's ills.

"You cannot promote trade at all costs because you will destroy the planet,' says Remi Parmentier, political director of Greenpeace International. "We've got a real fight over globalization going on. ❖

Reprinted with permission from **E/The Environment Magazine** Subscription Department: P.O. Box 2047, Marion, OH 43306 Telephone: (815) 734-1242 (Subscriptions are $20 per year) On the Internet: www.emagazine.com email: info@emagazine.com

Questions

1. What was held in Johannesburg, South Africa on August 26 through September 4?

2. How does the Kyoto Treaty get to the heart of consumption issues?

3. Why did the focus on the free market worry many environmentalists?

Answers are at the back of the book.

4 Over 5,000 people per day are forced to flee their homes because of environmental degradation and climate change. By the year 2050, these environmental refugees could number up to 150,000,000 people. Flooding, oil spills, rising sea levels, and a plethora of other causes are displacing a rapidly growing segment of the earth's population. In spite of the massive nature of this problem, world politicians and the UN continue to ignore this pressing issue.

Environmental Refugees

Mark Townsend

The Ecologist, July/August 2002

MARAT FOMENKO CASTS ONE FINAL GAZE over the bleak landscape. It is littered with fragments of abandoned machinery and the rusted hulks of disused ships.

Across the plain is Kazakhstan's once famous fishing port of Aralsk, and, beyond that, a huge, dried rubbish strewn sand pit.

It has been 25 years since Marat could see the receding Aral Sea once the lifeblood of the region and the fourth biggest lake in the world from his home.

The former fisherman motions to his wife Malika and their kids. The bags are packed. It is time to finally escape.

Marat's livelihood literally drained away from the moment the rivers that fed the Aral Sea were diverted to irrigate the pesticide-soaked cotton fields upstream in Uzbekistan.

The Fomenko family are heading to the Kazakh capital of Astana, joining the throng in search of a better life. They will never return to their dying homeland, the vanishing sea of which has triggered ecological disaster and a 30-fold increase in disease.

His son has contracted tuberculosis and Marat hopes the city will offer improved facilities.

Hope is all the Fomenko family have. They and 25 million others worldwide who have been forced to forever abandon their lands through a complex myriad of causes involving flooding, drought, soil erosion, deforestation, earthquakes, nuclear accidents and toxic spills.

These are the planet's environmental refugees. You may not have heard of them. Certainly, they are ignored by the world's politicians. And yet, experts argue, this rapidly swelling band of disparate, disenfranchised and displaced families constitutes one of the biggest crises facing humankind.

They are a huge, forgotten army of people whose numbers, according to conservative estimates, soar by 5,000 a day. Yet they are shunned by the international community, whose policies ensure they are deprived of not only basic rights, but actual recognition.

All corners of the globe are affected. There are vast swathes of land where the environment has become so degraded it can no longer support life. Each region of the world experiences its own specific agonies.

Just over 4,000 miles south of the shrinking Aral, Big Business is playing its part in this unfolding catastrophe.

Deep in Nigeria's Niger Delta lies the deserted home of Karalolo Atu. Three years ago Karalolo was forced from her ancestral kingdom of Ogoniland, and, along with thousands of others, she fled to the nearby settlement of Port Harcourt.

Quite simply Karalolo's local environment had collapsed. An alliance between oil giant Shell and corrupt, violent regimes had fuelled a complete breakdown of the fragile delta ecosystem.

Shell's unswerving search for fresh oil reserves had led to hundreds of oil spills.

Water systems and soil were left heavily polluted, and precious farmland was rendered unusable. Mother of four Karalolo cannot contemplate going back to her ruined homeland. Although millions of environmental refugees are displaced within the same country, the vast majority never return home because in most cases nothing is or can be done to reverse the damage.

Climate change

But the growing nightmare that will transform the surge of environmental refugees into a problem of unimaginable dimensions is, unquestionably, climate change.

Just ask farmer Paani Talake from the tiny island state of Tuvalu in the South Pacific. His thatched family home is literally going down in history.

Whereas Marat's problem is a shrinking sea, Paani's is altogether different. For the latter there will soon be nothing left but sea.

Already the lowland coconut plantation farmlands of Tuvalu are being swamped by the rising sea. Nearby islets have vanished forever, while the invisible creep of saltwater contaminates precious drinking supplies and stunts crop growth. Next year Paani and his young family will abandon their homeland and take advantage of a gracious offer of a new start from the government of New Zealand.

Paani has little choice. Within as little as 50 years Tuvalu is projected to slide beneath the encroaching waters—a high-profile victim of the industrial excesses of the West. All that will be left of Tuvalu will be its status as a graphic footnote to mankind's folly in experimenting with the atmosphere.

But what of the millions of others in low-lying countries who may soon join the flow of environmental refugees? Where will they be offered a new start?

Under official predictions, their islands and coastlines will soon start sliding into the rising tide as climate change propels the planet into a new stratosphere of catastrophe. Ever greater numbers will be forced to scratch harder for a living on less and less land—land which is already struggling to sustain current demands.

The Intergovernmental Panel on Climate Change's forecast of a one-metre sea-level rise this century poses one of the largest dilemmas yet to face the human race. The prospect is particularly bleak given the fact that half the planet's people are already crowded into coastal zones. Some 10 million of these people are at constant risk of flooding.

In Bangladesh alone a one-metre rise would uproot 20 million people. Then there are the vast rice-growing river floodplains of Thailand, Indonesia and India, among others.

Even the rich world must pay a price. There are devastating implications for nations such as Holland and Denmark, with the possibility of huge population shifts and waves of environmental refugees moving onto already cramped lands.

Such massive migration will be accompanied by the stench of sickness. Mosquito-borne diseases are expected to increase 100-fold in temperate regions. Malaria has already quadrupled in the last five years.

Incredibly, politicians have chosen to ignore the impending crisis, refusing to accept the likes of Paani, Marat and Karalolo are the refugees they most unequivocally are.

Even the United Nations High Commissioner for Refugees (UNHCR), established in 1950 in response to the mass-displacement of Europeans in WWII, has conspicuously failed to address the problem. The UNHCR refuses to update its legal framework in line with the planet's rapidly deteriorating environment. As for a tangible solution, forget it.

Instead the agency clings onto the politically narrow, outdated definition of refugees, which stipulates that people should only be considered as such if their flight is due to 'a well-founded fear of persecution' on grounds such as race and religion.

But doesn't an environment which has become so degraded that it no longer offers the basic building blocks of life—namely, food and water—persecute?

The upshot is that at least 25 million refugees (though the true total is likely to be far higher) are not afforded basic rights.

These people's plight was similarly glossed over by the UK's Refugee Week in June. Refugee Week preferred to concentrate on Britain's asylum seekers. It overlooked figures from the Red Cross, which show that more than half—58 per cent—of the world's 43 million refugees are in fact environmentally displaced. In other words, almost one in every 250 persons on our planet.

All future trends point to an acute escalation of environmentally-driven human migration. Dr Norman Myers, a visiting fellow at Oxford University, believes that climate change and environmental degradation will create 150 million environmental refugees by 2050.

Klaus Topfer, chief executive of the United Nations Environment Programme, says that the swollen ranks of environmental refugees could double to 50 million in just eight years time. That is an increase of 8,500 a day.

But even Topfer may as well be whispering in the wind.

The Politics of Environmental Refugees

Asylum is a topic that carries the power to make and break politicians. Just look at the improbable success of Jean-Marie Le Pen in France. But despite asylum issues dominating both the media and politics, the actual role of the world's degraded environment as a factor in human migration is being conveniently ignored. Thus politicians and media magnates can continue labeling environmental refugees as 'bogus'. So, refugee policy is concocted in seeming oblivion to the problem of environmental refugees, and Western governments are allowed to act as if loss of millions of legitimate migrants do not exist.

That is why the British Home Secretary David Blunkett's recently published White Paper on immigration and asylum failed to concern itself with environmental refugees. Blunkett preferred to pander to the whims of Fortress Europe instead. A mature analysis into why people are migrating in the first place never takes place.

It is almost as if the unfolding problem is too big to comprehend. It is easier and cheaper to ditch the 3,500-year-old tradition of affording succor to refuges, and to systematically deny the likes of the Fomenko family the right to a better life.

Wendy Williams, population movement advisor for the International Red Cross, is under no illusion that politicians are purposefully avoiding the repercussions of environmental collapse in order to keep numbers of 'legitimate' refugees down.

"If politicians relaxed migrations laws," Williams says, "it would probably be their death knell. We need to raise awareness that these people simply cannot survive off the land anymore, and that they don't want to leave their homes in the first place."

For all the bluster, Britain's 'refugee crisis' remains piffling compared to the size of the true environmental problem.

If global trends for environmental refugees were applied to the UK, there would be around 250,000 people—the equivalent of the population of Sunderland—thus affected in this country. That is three times the record number of asylum applicants for a single year in Britain.

Green MEP for London Jean Lambert is one of the few politicians in the West who admits to being intensely worried about the problem. In May she unveiled a detailed report into the environmental refugee crisis. The report outlined her concerns that a serious debate has yet to commence on the unfolding crisis.

Lambert is flabbergasted that the issue of finding new homes for 10s of millions of people in the near future is not even worthy of peripheral concern.

The UNHCR

Just how low a priority the issue is can be illustrated by the fact that the annual budget of the UNHCR is a mere 843m. That is less than the military expenditures of world governments in a single day. It is arguably barely enough to cope with the demands of conventionally 'persecuted' migrants.

Of the UNHCR money, a fraction so tiny it cannot be easily broken down is offered to environmental refugees.

There are no publicised plans to increase help to these migrants, even though they constitute the majority of the world's displaced. The British government alone spends almost the same amount—835m—handling asylum seekers in this country.

The irony is that the developing world continues to be hit hardest by environmental degradation and human-driven climate change. That suffering seems to be in direct disproportion to the developing world's responsibility for climate change. After all, the US alone spews out 25 per cent of greenhouse gases on behalf of just four per cent of the world's population.

From the terrorised perspective of the Paani family's thatched roof, the US's refusal to cooperate with the Kyoto Protocol must seem grotesquely indifferent to say the least.

Driven From Their Land

Even when climate change is removed from the frame the picture remains grim. Soaring population growth and devastated, exhausted environments are creating immense suffering and massive migration on their own.

A whistle-stop tour of the world makes disquieting reading. Mexico, the Ivory Coast and the Phillipines could all lose the bulk of their forests within half a lifetime. In the same short timescale Ethiopia, El Salvador and Nepal could lose most of their farmland topsoil.

Globally, one in three people face acute water shortages as water use is expected to increase by 40 per cent over the next 20 years. Many of these people will be forced from their homes to seek clean water supplies elsewhere. Countries like Jordan, Egypt and Pakistan will be particularly affected. India's breadbasket—the huge agricultural plains of the Punjab—is already more than half eroded.

And almost overnight an abundance of land in countries like Kenya and Costa Rica has been dramatically transformed into acute land shortage through rapid urbanisation.

Each of these factors independently could trigger extraordinary numbers of environmental refugees. And all the time the pressures are growing.

In the 15 minutes it takes you to read this article, the world will gain another 2,600 mouths to feed; 97 of every 100 will be born into a country where finances are stretched, food and water is insufficient and where creaking, chaotic cities are groaning under the weight of incoming migrants. These cities are mostly poised on the brink of natural disaster: 40 of the world's 50 fastest-growing cities are stranded within earthquake zones.

Once again we come back to climate change. Four years ago the world saw the birth of the 'super-disaster'. For the first time in history more people were being displaced because of environmental reasons than war. A report from the International Red Cross and Red Crescent Societies warned that the number of people they had helped after major floods, droughts and earthquakes had increased from 500,000 to five and a half million in just six years. A UN survey estimates that around a third of the world's total land is in the process of becoming infertile. While massive man-made projects like China's Three Gorges dam are driving more than one million people from their homes. Other studies predict that 100 million of 135 million people living in areas of desertification will be displaced in the next 20 years.

These are just some of the complex, alarming web of factors powering this new wave of refugees. It is a complexity that will prove taxing for politicians. But, unless they start attempting to solve it, it will store up even greater problems for the future.

Solutions?

Measures need to be introduced to ensure Paani's fate is not repeated across the world.

Putting the brakes on climate change will only be achieved by reducing greenhouse emissions by 90 per cent (not 10 or 20 per cent) within a decade.

One interesting development that could hold huge ramifications for Western governments is the threat by Paani's prime minister to take legal action against polluter states for greenhouse gas emissions.

But first we need a definition of refugees that includes those displaced for environmental reasons.

Redefining state responsibility for environmental refugees is another must—a tough choice for leaders who must start reacting to the fact that people are being pushed from their homes and not pulled by the bright lights of the West.

Running out of time

The problem is growing daily. An action plan is needed. Finding new homes for 125 million people in a few decades will test even the most committed.

As its stands the world is not prepared to deal with these implications. It is barely aware of the impending crisis. Disaster awaits, only a dramatic upsurge in political will can prevent tens of millions of people from experiencing the same desperate fate as the families of Paani, Matra and Karalolo. ❖

Drowning By Numbers: The Consequences of Rising Sea Levels and Subsidence

- In China, the city of Shanghai could be entirely flooded. The government calculates that 30 million of its people could be displaced by global warming.
- With a forecast of 142 million people inhabiting coastal India by 2050, India's flood-zone refugees could total anything between 20 and 60 million.
- Seven per cent of Bangladesh could be permanently lost, with an estimated 15 million people being displaced.
- By 2050 Egypt is expected to lose between 12 and 15 per cent of its arable land, with a possible 14 million people being displaced. Egypt already imports well over half its food.
- Other delta areas at risk include Indonesia, Thailand, Mozambique, Gambia and Senegal.
- Island states at risk include the Maldives, Kiribati, the Marshalls and dozens of Caribbean states. Around 1 million people are likely to have to evacuate permanently.

Source: N. Myers, 'Environmental Refugees in a Globally Warmed World', BioScience, Vol 43/11, December 1993.

The article first appeared in the July/August 2002 issue of the *Ecologist* (Volume 32, No. 6) www.theecologist.org.

Questions

1. What has forced 25 million people worldwide to abandon their lands?
2. According to Dr. Norman Myers, how many refugees will climate change and environmental degradation create by 2050?
3. Putting the brakes on climate change will only be achieved by reducing greenhouse emissions by what percent?

Answers are at the back of the book.

Environment of Life on Earth

5

Ecologists have long known that the small, inconspicuous species in any ecosystem are far more important to the health and stability of that ecosystem than most people realize. Thus, while the general public might not notice that a small herring, called menhaden, has declined over 50% in the last decade, marine ecologists are not surprised to learn that this decline may well signal a catastrophic loss to the food web of the ocean. The loss of this "keystone species" could upset the balance of nature in coastal waters for many decades to come.

The Most Important Fish in the Sea

H. Bruce Franklin

Discover, September 2001

FIRST YOU SEE THE BIRDS—gulls, terns, cormorants, and ospreys wheeling overhead, then swooping down into a wide expanse of water dimpled as though by large raindrops. Silvery flashes and splashes erupt from thousands of small herringlike fish called menhaden. More birds arrive, and the air rings with shrill cries. The birds alert nearby anglers that a massive school of menhaden is under attack by bluefish.

The razor-toothed blues tear at the menhaden like piranhas in a killing frenzy, gorging themselves, some killing even when they are too full to eat, some vomiting so they can kill and eat again. Beneath the blues, weakfish begin to circle, snaring the detritus of the carnage. Farther below, giant striped bass gobble chunks that get by the weakfish. From time to time a bass muscles its way up through the blues to take in whole menhaden. On the seafloor, scavenging crabs feast on leftovers.

The school of menhaden survives and swims on, its losses dwarfed in plenitude. But a greater danger than bluefish lurks nearby. The birds have attracted a spotter-plane pilot who works for Omega Protein, a $100 million fishing corporation devoted entirely to catching menhaden. As the pilot approaches, he sees the school as a neatly defined silver-purple mass the size of a football field and perhaps 100 feet deep. He radios to a nearby 170-foot-long factory ship, whose crew maneuvers close enough to launch two 40-foot-long boats. The pilot directs the boats' crews as they deploy a purse seine, a gigantic net. Before long, the two boats have trapped the entire school. As the fish strike the net, they thrash frantically, making a wall of white froth that marks the net's circumference. The factory ship pulls alongside, pumps the fish into its refrigerated hold, and heads off to unload them at an Omega plant in Virginia.

Not one of these fish is destined for a supermarket, canning factory, or restaurant. Menhaden are oily and foul and packed with tiny bones. No one eats them. Yet they are the most important fish caught along the Atlantic and Gulf coasts, exceeding the tonnage of all other species combined. These kibble of the sea fetch only about 10 cents a pound at the dock, but they can be ground up, dried, and formed into another kind of kibble for land animals, a high-protein feed for chickens, pigs, and cattle. Pop some barbecued wings into your mouth, and at least part of what you're eating was once menhaden.

19

Humans eat menhaden in other forms too. Menhaden are a key dietary component for a wide variety of fish, including bass, mackerel, cod, bonito, swordfish, bluefish, and tuna. The 19th-century ichthyologist G. Brown Goode exaggerated only slightly when declaring that people who dine on Atlantic saltwater fish are eating "nothing but menhaden."

And that is one problem with the intensive fishing of menhaden, which has escalated in recent decades. This vital biolink in a food chain that extends from tiny plankton to the dinner tables of many Americans appears to be threatened. The population of menhaden has been so depleted in estuaries and bays up and down the Eastern Seaboard that even marine biologists who look kindly on commercial fishing are alarmed. "Menhaden are an incredibly important link for the entire Atlantic coast," says Jim Uphoff, the stock assessment coordinator for the Fisheries Service of the Maryland Department of Natural Resources. "And you have a crashing menhaden population with the potential to cause a major ecosystem problem." Menhaden have an even more important role that extends beyond the food chain: They are filter feeders that consume phytoplankton, thus controlling the growth of algae in coastal waters. As the population of menhaden declines, algal blooms have proliferated, transforming some inshore waters into dead zones.

To grasp how ubiquitous menhaden once were, you can read the journals of explorer John Smith. In 1607, he sailed across the Chesapeake Bay through a mass of menhaden he described as "lying so thick with their heads above the water, as for want of nets (our barge driving amongst them) we attempted to catch them with a frying pan." Colossal schools of menhaden, often more than a mile in diameter, were once common along the entire Atlantic and Gulf coasts of the United States. Since World War II, however, fishermen using spotter planes and purse seines appear to have dramatically decreased both the population and the range of menhaden.

Bryan Taplin, an environmental scientist in the Atlantic Ecology Division of the Environmental Protection Agency (EPA), has witnessed the destruction of all the large schools of menhaden by purse seiners in Rhode Island's Narragansett Bay. During the last two decades he has also studied changes in the diet of striped bass in the bay by analyzing the carbon isotope signature of their scales. What he has discovered is a steady shift away from fat-rich menhaden to invertebrates that provide considerably lower nutritional value. That has been accompanied by a loss of muscle and a decrease in the weight-to-length ratio of striped bass. The bass that remain in Narragansett Bay, says Taplin, are "long skinny stripers" that have been forced to shift their diet because "the menhaden population has crashed to an all-time low."

"You have to scratch your head and wonder—since we set quotas for bluefin and tuna—why we don't set quotas for this crucial part of the oceanic food chain," says Taplin. "Not to regulate a fishery that's so important is to ask for trouble. I wonder whether we are about to see something go wrong unlike anything we have ever seen."

Signs of what could go wrong are already obvious in the Chesapeake Bay, the tidal estuary that once produced more seafood per acre than any body of water on Earth. "There's nothing in Chesapeake Bay that can take the place of menhaden," says Uphoff of the Maryland Fisheries Service. "Menhaden are king." Jim Price is a fifth-generation Chesapeake Bay fisherman. For 10 years he captained a charter boat specializing in light-tackle fishing for striped bass, also called rockfish by bay anglers. One day in the fall of 1997, Price caught a rockfish so diseased he still becomes upset when he talks about it. "I'd never seen anything like that in my entire life," he says, wringing his powerful, deeply tanned hands. "It was covered with red sores. It was so sickening it really took something out of me."

Price deposited several sick rockfish at the Cooperative Oxford Laboratory in nearby Oxford, Maryland, and then began his own independent study. When he cut some open, he was shocked. "I've been looking in the stomachs of rockfish for 40 years," he says, "but I couldn't believe what I saw—nothing, absolutely nothing. Not only was there no food, but there was no fat. Everything was shrunk up and small."

An Oxford lab pathologist speculated that the fish might have been "decoupled from their source of food," but Price was incredulous. "I thought to myself, with all the food here in the Chesapeake, that's a stupid idea. Then I got to thinking. In years past, at that time of year I would find their stomachs full of menhaden, sometimes a half-dozen whole fish."

Price hypothesized that malnutrition, caused by the decline in the menhaden population, made the rockfish vulnerable to disease. Since then, his hypothesis has been confirmed by research. Half the rockfish in the Chesapeake are diseased, with either bacterial infections or lesions associated with *Pfiesteria*, a toxic form of phytoplankton known as the cell from hell. But that is only one symptom of the depletion of menhaden.

Dense schools of menhaden swimming with their mouths open slurp up enormous quantities of plankton and detritus like gargantuan vacuum cleaners. In the Chesapeake and other coastal waterways, the filtering clarifies water by purging suspended particles that cause turbidity, allowing sunlight to penetrate to greater depths. That encourages the growth of plants that release dissolved oxygen as they photosynthesize. The plants also harbor fish and shellfish.

Far more important, the menhaden's filter feeding limits the spread of devastating algal blooms. Runoff from many sources—farms, detergent-laden wastewater, overfertilized golf courses, and suburban lawns—floods nitrogen and phosphorus into coastal waters. Nitrogen and phosphorus in turn stimulate the growth of algal blooms that block sunlight and kill fish. The blooms eventually sink in thick carpets to the sea bottom, where they suck dissolved oxygen from the water and leave dead zones. Menhaden, by consuming nutrient-rich phytoplankton and then either swimming out to sea in seasonal migrations or being consumed by fish, birds, and marine mammals, remove a significant percentage of the excess nitrogen and phosphorus that cause algal overgrowth.

Nature had developed a marvelous method for keeping bays and estuaries clear, clean, balanced, and healthy: Oysters, the other great filter feeders, removed plankton in lower water layers, and menhaden removed it from upper layers. As oysters have been driven to near extinction along parts of the Atlantic coast, menhaden have become increasingly important as filters.

Marine biologist Sara Gottlieb says: "Think of menhaden as the liver of a bay. Just as your body needs its liver to filter out toxins, ecosystems also need those natural filters." Overfishing of menhaden is "just like removing your liver," she says, and "you can't survive without a liver."

During the late 19th century, several dozen sailing vessels and a handful of steamships hunted menhaden in Gardiners Bay, near the eastern tip of Long Island, New York. The abundance of menhaden then appealed to another set of hunters: ospreys that nested in an immense rookery on Gardiners Island. As late as the mid-1940s, there were still 300 active osprey nests on the small island. But the ospreys fell victim to the DDT that was sprayed on the wetlands. Eventually, the number of active nests plummeted to 26. After DDT was banned, biologist Paul Spitzer observed a gradual resurgence of the osprey. However, in recent years he has watched the number of ospreys on Gardiners Island dwindle again. From 1995 to 2001, he says, "there has been an absolute steep decline from 71 active nests to 36."

Although no longer weakened by toxins, ospreys now have little to eat. "Migratory menhaden schools formerly arrived in May, in time to feed nestlings," Spitzer says. In recent years, menhaden have disappeared, and the survival rate of osprey chicks has fallen to one chick for every two nests, a rate comparable to the worst years of DDT use. "The collapse of the menhaden means the endgame for Gardiners Island ospreys," he says. Spitzer sees the same pattern of decline in other famous osprey colonies, including those at Plum Island, Massachusetts; Cape Henlopen, Delaware; Smith

Point, New York; and Sandy Hook and Cape May in New Jersey.

The menhaden crash may also contribute to the decline of the loons that make an autumn migration stopover in the Chesapeake each year. Spitzer keeps statistical counts of flocks passing through a roughly 60-square-mile prime habitat on the Chesapeake's Choptank River, near where Jim Price found diseased striped bass. Between 1989 and 1999, Spitzer's loon count dropped steadily from 750 to 1,000 per three-hour observation period to 75 to 200. The typical flock fell from 100 to 500 birds to between 15 and 40. Menhaden are "the absolute keystone species for the health of the entire Atlantic ecosystem," says Spitzer.

Hall Watters, now 76 and retired, looks back ruefully on the role he and other spotter pilots played in the demise of the menhaden. "We are what destroyed the fishery, because the menhaden had no place to hide," he says. "If you took the airplanes away from the fleet, the fish would come back."

Watters was the first menhaden spotter pilot, hired in 1946 by Brunswick Navigation of Southport, North Carolina. He had been a fighter pilot during World War II and says he was "the only pilot around who knew what menhaden looked like." Brunswick had just converted three oceangoing minesweepers and two submarine chasers to menhaden fishing ships and was eager to extend the range and efficiency of its operations. Menhaden usually spawn far out at sea, and the larvae must be carried by currents to the inshore waterways where they mature. Guided by Watters, Brunswick's rugged vessels soon began to net schools as far out as 50 miles, some with so many egg-filled females, he says, that the nets "would be all slimy from the roe."

Watters remembers that in the early postwar years, menhaden filled the seas. In 1947, he spotted one school about 15 miles off Cape Hatteras so large that from an altitude of 10,000 feet, it looked like an island. Although 100 boats circled the school, many fish escaped. "Back then we only fished the big schools. We used to stop when the schools broke up into small pods." But things had changed dramatically by the time he quit in 1980: "We caught everything we saw. The companies wanted to catch everything but the wiggle."

The exact size of the Atlantic menhaden population in 2001 is impossible to measure, but industry statistics show a dramatic decline in catches over the years since 1946. The average annual tonnage from 1996 to 1999 was only 40 percent of the average annual tonnage caught between 1955 and 1961. Last year the catch was the second lowest in 60 years. Moreover, these numbers may not reflect the full scope of the decline because the catch is not necessarily proportional to the population. "The stock gets smaller but still tends to school," says Jim Uphoff of the Maryland Fisheries Service. "The fishery gets more efficient at finding the schools. Thus they take a larger fraction of the population as the stock is going down."

The large oceanic schools of menhaden are often too scarce to chase profitably, so the fishing industry has moved into estuaries and bays, particularly the Chesapeake. Maryland has banned purse seining in its portion of the Chesapeake. Virginia has not. Omega Protein, headquartered in Houston and the largest U.S. menhaden fishing firm, has almost unlimited access to state waters, including the mouth and southern half of the Chesapeake. By 1999, 60 percent of the entire Atlantic menhaden catch came from the Virginia waters of the Chesapeake.

These days Omega Protein enjoys a near monopoly fishing for menhaden. As the fish population declined and operational costs increased, many companies went bankrupt or were bought out by bigger, more industrialized corporations. Omega Protein's parent was Zapata, a Houston-based corporation cofounded by former president George Bush in 1953. Omega Protein went independent in 1998, after completing the consolidation of the menhaden industry by taking over its large Atlantic competitor, American Protein of Virginia, and its Gulf competitor, Gulf Protein of Louisiana.

Omega Protein mothballed 13 of its 53 ships last year and grounded 12 of its 45 spotter planes as the menhaden continued to disappear. Fewer than a dozen of the company's ships fish out of Virginia, but 30 ships fish the Gulf of Mexico.

The Gulf seems to be headed for the same problems that are obvious in the Chesapeake, but on a larger scale. Fed by chemical runoff, algal blooms have spread, causing ever-enlarging, oxygen-depleted dead zones. And jellyfish are proliferating, both a native species and a gigantic Pacific species. Researchers believe the swollen jellyfish population could have a devastating effect on Gulf fishing because they attack the eggs and larvae of many species. Monty Graham, senior marine scientist at the Dauphin Island Sea Lab in Alabama, says overfishing, "including aggressive menhaden fishing," seems to have allowed the jellyfish—"an opportunistic planktivore"—to fill the ecological void. He says the proliferation of both species of jellyfish indicates "something gone wrong with the ecology."

Barney White, corporate vice president of Omega Protein and chairman of the National Fish Meal and Oil Association, the industry's trade association, categorically denies that menhaden are being overfished or that there is any ecological problem whatsoever caused by their decline. He says the controversy "is largely without basis" and is based on "lies" disseminated by recreational fishermen in general and Jim Price in particular. "It becomes an issue of politics rather than science—that people have a problem with commercial fishing in general," White says. "We have big boats closer to shore, so we're easy to see, and that makes us a convenient political target."

White attributes the absence of adult fish in New England and eastern Long Island waters to cyclic factors. "Well-meaning people who don't know marine biology have been mistaking short-term occurrences for long-term trends," he says. "In fact, the reports I have show that more fish seem to be moving into the area." Moreover, White says, "the total biomass is sufficient to sustain the industry."

Watters disagrees. More than a half century after he first took to the air as a spotter pilot, he fumes that "the industry destroyed their own fishery, and they're still at it." What galls him the most is that an increasing proportion of the catch consists of "zeros"—menhaden less than a year old. He advocates banning menhaden fishing close to shore, especially in estuaries, where the young menhaden mature. He also argues that if Omega Protein "enlarged the mesh size, they wouldn't be wiping out the zero class."

White acknowledges the industry is facing a problem of "recruitment"—menhaden are not living through their first year. But he insists that the 13/4-inch mesh now used allows the very smallest juveniles to slip through. The real problem, he says, is "an overpopulation of striped bass. We think the striped bass are eating all the juveniles."

Omega Protein's financial reports indicate that the fortunes of the company rise and fall with "the supply and demand for competing products, particularly soybean meal for its fish meal products and vegetable oils and fats for its fish oil products." The fishing industry's journal, *National Fisherman*, says: "On the industrial side of the fishery, where menhaden is processed into feed for poultry and pigs, the demand for fish is depressed by a surplus of soy, which serves the same purpose." In other words, all the ground-up menhaden could be replaced by ground-up soybeans.

Since market forces are unlikely to curtail the menhaden fishery, governments may have to take action. Price thinks the fishing season for menhaden should be closed each December 1, "because after that is when the age zeros migrate down the coast." No matter what is done, most researchers agree the menhaden must be viewed not as a specific problem about a single species of disappearing fish but as a much larger ecological threat.

Bill Matuszeski, former executive director of the National Marine Fisheries Service and former director of the EPA's Chesapeake Bay program, says: "We need to start managing menhaden for their role in the overall ecological system. If this problem isn't taken care of, the EPA will have to get into the decision making." Matuszeski believes estuaries like the Chesapeake Bay should be put off limits to menhaden fishing immediately. "That would be inconvenient for the industry, but it would be inconvenient for the species to be extinct." ❖

Questions

1. Menhaden make up what percentage of the catch of commercial fisheries in the United States?
2. What can menhaden be used for?
3. When did the menhaden population begin to decline?

Answers are at the back of the book.

6

Trout species are increasingly being threatened by illegal stockpiling and other activities. While some measures are in place to protect trout species, the dangers facing North America's trout are largely ignored or resisted by anglers and environmentalists alike. The importance of trout to North American ecosystems and successful trout protection programs are examined.

Trout Are Wildlife, Too

Ted Williams

Audubon, December 2002

On August 6, 2002, the PMDs started coming off Armstrong Spring Creek at 10:00 a.m. PMDs (pale morning duns) are delicate yellow mayflies that shuck their larval skins on the surface and, if they don't vanish into the maws of trout, dance around like garden fairies, with sunlight flashing on translucent wings. A spring creek leaps full grown from rocks or wet earth. Armstrong, the most famous spring creek in the world, is collected by the Yellowstone River in Montana's Paradise Valley.

As I stood in the icy flow, nighthawks and swallows dipped from the cloudless, mountain-rimmed sky, picking off emerging PMDs, while all around me large trout were finding plenty of their own, bulging through the surface and wagging flaglike dorsal fins. These were browns and rainbows that, at the sting of my hook, somersaulted into the air and raced off on long runs, most of which ended with a sickening snap because my fluorocarbon "tippet" was so fine it broke at 1.5 pounds of pressure. Anything heavier and the trout would refuse my PMD imitation. As an angler I've been trained to measure the quality of game fish by this kind of strength and selective feeding behavior. But as a naturalist I'm conflicted. Browns evolved in Europe, rainbows in the Pacific

Northwest. Both were unleashed decades ago in the interior West by managers blind to the beauty and importance of native ecosystems.

The trout that belong here are cutthroats. Rainbows hybridize with them, swamping their genes. Browns displace them, as do brook trout (imported from the East). Of the 14 named and unnamed cutthroat subspecies, two are already extinct, and the rest are in desperate trouble, pushed into river tops where they're protected from alien invaders by waterfalls or manmade barriers but where they're also genetically isolated.

Late in the day, when the PMDs were gone, I was delighted and astonished to catch a Yellowstone cutthroat, the native subspecies of this river system. It slurped my beetle pattern on a sloppy drift, and it came in easily, shaking its head and rolling. All wild trout are beautiful, but cutthroats mesmerize me. This one glowed with the gold of autumn aspens and the pinks of a Big Sky sunset. Its flanks were flecked with obsidian spots that got bigger and more profuse toward the tail, and under its jaw were the two scarlet slashes that give the species its name. Cutthroats are hardwired: They're not selective, because they evolved in sterile water where they couldn't afford to let something drift by that might have been a bug; and they never developed the kind of energy-draining musculature of other trout. When the state of Idaho sought to restore Yellowstone cutts to Island Park Reservoir, one prominent guide—an educator of local anglers—declared: "They're stupid, and they fight like slugs." So fierce was public opposition that the project was abandoned.

In the Yellowstone drainage, however, cutthroats are making a comeback, because trout managers of the Montana Department of Fish, Wildlife and Parks are the most progressive in the nation. They've leased water rights on tributaries dewatered by irrigators. Now native trout are spawning in these rivers again.

Montana has learned that hatcheries, which the angling public underwrites with license fees and a federal tax on fishing tackle, are among the greatest threats to wild (stream-bred) trout, whether naturalized or native. Genetic diversity, by which trout adapt to different habitats in large river systems, is bred out of hatchery trout. They are selected for domesticity, warped by inbreeding. They survive in the real world only long enough to suppress and displace wild trout. Moreover, hatcheries spread pathogens such as whirling disease, imported from Europe with frozen pike and to which North American trout lack natural immunity. But when game and fish departments try to phase out hatcheries, anglers—unwilling to learn the truth—scream to their legislators, who threaten budget cuts. "If you cross a sacred cow with a military base, you get a fish hatchery," says Bernard Shanks, the gutsy former director of the Washington Fish and Wildlife Department, who tried to de-emphasize hatchery production.

In 1970 Montana stopped stocking hatchery fish (browns and rainbows) in a section of the Madison River where these species had long been established and which is far too big for native-trout restoration. Four years later large fish (three years and older) were up 942 percent. The study horrified anglers and hatchery bureaucrats, who wanted to believe that stocking was the key to trout abundance. In apparent sabotage, the study area was stocked in 1972 (presumably with trout purchased at a private hatchery) and the towing hitch on the department's truck was loosened so that boat and trailer parted company on the highway. The illegal stocking only corroborated the earlier data, because immediately the brown trout biomass dipped by 24 percent, then, with two more years of no stocking, jumped back to where it had been. The study convinced Montana to cease all trout stocking in moving water. As a result it is now the number-one trout-fishing destination in the nation.

In most habitats, trout are not easily seen. Except where they are conditioned with pellets, they don't come to feeders.

While they're every bit as colorful as birds, they're cold and slimy, and most of the public remains unmoved by their plight. Groups such as Trout Unlimited and the Federation of Fly Fishers are winning important battles for native-trout restoration, but they're outnumbered and outshouted by the gull-like masses for whom trout genes (and even trout fins, abraded into fleshy stumps by the sides of hatchery raceways) have no relevance, for whom a trout is not part of a native ecosystem but a slab of meat.

Anyone seeking the answer to "What good is a native trout?" need not look beyond Yellowstone National Park. Eighty percent of the world's remaining pure Yellowstone cutthroats abide in 87,000-acre Yellowstone Lake, spawning in at least 59 feeder streams. Today Yellowstone cutts fuel aquatic and terrestrial ecosystems in and around the park the way sockeye salmon fuel ecosystems in southern Alaska. But it wasn't always this way. Thirty years ago the park's native trout had been pretty much wiped out. Dead cutthroats—caught, killed, and discarded by tourists—comprised the main item in park garbage cans. Grizzlies, sustained by this and other garbage, had been reduced to circus bears. Then, in the early 1970s, a smart, tough biologist named John Varley (who now directs the park's Center for Resources) led a successful effort to require anglers to release most of their trout. The initiative was far more contentious than wolf reintroduction (which Varley also led). Outfitters charged the park with plotting to "put them out of business." Outdoor writers reported that the feds planned to end all sportfishing. Fisheries managers parroted the old wives' tale that "you can't stockpile trout."

When the Craighead brothers were studying grizzlies in the 1960s, they never saw a bear take a fish. By 1975 bear activity was being observed on 17 of the lake's 59 cutthroat-spawning streams. Now bears work at least 55 of those streams, and one research team has observed a sow with cubs averaging 100 fish a day for 10 days. In 1988 there were 66 nesting pairs of ospreys in the park; by 1993 there were 100. While in the park, white pelicans get almost all their nourishment from cutthroats, consuming an estimated 300,000 pounds a season. In all, Yellowstone cutts provide an important food source for at least 28 species of birds and mammals.

Last July 18 I stood in the swollen Yellowstone River in the park's Hayden Valley with the current piling up around the top of my chest waders. I fish here not to "fight" native trout but to connect with them and their world. Sometimes the insect hatches are so prolific that the fish don't bother to rise; they just hold in the current with their heads out of water and, like drunks under wine spigots, let the river's richness fill their bellies. The nutrient flow starts with the sulfurous fumes that bubble white and pungent from underwater vents; cycles skyward with squalls of caddises and mayflies; drops back to the trout; then out onto the banks with the otters and minks; up and south with the eagles and ospreys; seaward with the loons and pelicans; high into the stream-etched Absarokas with the massive spawning run; and, finally, into the gullets of grizzlies.

The cutthroats of Yellowstone Lake, restored by the no-kill fishing regs that were going to ruin the outfitters, now generate about $36 million a year for them and other local businesses, and the figure doubles when you include other restored park waters. But now Yellowstone's trout-based ecosystem and trout-based economy may collapse again. Lake trout, unavailable to wildlife because they live and spawn in deep water, have been illegally stocked in the lake, and wherever these large, voracious predators have been superimposed on native cutthroats, the cutthroats have been eliminated.

When a fisherman caught the first lake trout on July 30, 1994, the news made Varley physically ill. Unless the park can permanently suppress the aliens (elimination is out of the question), Yellowstone cutthroats are doomed in the lake and probably the world. In a never-ending project that leaves virtually no money for other trout restoration, crews on two large boats set gill nets at depths favored by lake trout. They're getting good at it; in 2001 they killed 15,000 lake trout with an accidental cutthroat bycatch of only 600. But there are alarming indicators. On Clear Creek the cutthroat spawning run has declined from about 12,000 to 8,000 fish; there's a corresponding decline at sample stations in the lake.

Particularly discouraging is the ignorance of sportsmen. Facing a future no less bleak than the Yellowstone cutthroat is the westslope cutthroat. Ambitious restoration projects are under way in Montana, where pure westslopes have been driven out of something like 75 percent of their historic range. But the most ambitious westslope-restoration project ever proposed has been derailed for the past three years by a sportsmen-endorsed property-rights group called the Public Lands Access Association. The group's president, Bill Fairhurst, threatened to sue the state in federal court on the grounds that it would "pollute" public water with the safe, selective, short-lived fish poisons with which it plans to remove the brook trout, rainbows, and hybrid cutthroats that infest 77 miles of Upper Cherry Creek, in southwest Montana. What's really bugging the association and its allies is that 85 percent of the project area is owned by media mogul and native-ecosystem champion Ted Turner, who has offered to pick up $343,350 of the $475,000 cost. Like the previous owner, Turner does-

n't invite the public onto his land, although the Montana access law permits anglers to wade Cherry Creek.

In January 2002 *Fly Rod & Reel* magazine, where I serve as conservation editor, recognized Turner's commitment to native trout by making him its Angler of the Year, thereby eliciting the biggest blizzard of nastygrams we've seen in our 23-year history. I had "a political agenda," I'd done it for money, I was a "snot nose," a "moron," a "nasty bully," a "nature Nazi," an acolyte of "Hanoi Jane," an espouser of "vitriolic leftist environmentalism." "I see your magazine is lining up lock-step with the wild-animal-rights fly-fishing crowd that Left Wing Ted [Turner] leads and which appears to be taking over the leadership of Trout Unlimited and the Federation of Fly Fishers . . ." wrote Bruce Cox of Springdale, Pennsylvania. "I am completely opposed to the wild-at-any-cost perspective of this left-wing animal-rights crowd and to wit will . . . politically align myself with anti-wild-fish groups and politicians."

Preserving Cherry Creek's alien and mongrel trout was the priority of most readers we heard from. The fishing was already good—why change species? Anglers had been programmed by the mass-circulation hook-and-bullet press, particularly *Outdoor Life* magazine, which had attacked the project with an article rife with misinformation entitled "Playing God on Cherry Creek." When the editors invited readers to vote for or against making Cherry Creek a sanctuary for westslope cutthroats, 98 percent voiced opposition.

In the late 1980s the Idaho Department of Fish and Game announced that it would cease polluting the Big Wood River with hatchery trout. But to appease the masses, which had threatened legislative intervention, the department kept stocking a few token fish. Idaho also went to wild-trout management on the Teton River but found it necessary to buy a four-acre gravel pit—safely isolated from the river—into whose seepage it poured a gravy train of hatchery fish. This direct dump-and-catch approach proved so popular that the department now does it all over the state.

On Henrys Lake, an important sanctuary for Yellowstone cutts, Idaho Fish and Game had been stocking rainbow-cutthroat hybrids because they fight harder. But in 1976, when managers announced they would stop stocking the lake with manmade mongrels, anglers threw a hissy fit and got the legislature to hold hostage the department's budget. So today the stocking of Henrys Lake continues, but with sterile "triploid" rainbows, which have three sets of chromosomes instead of the normal two and which hatchery technicians produce by heat-shocking the eggs. Despite the wastefulness and tastelessness of this strategy, native cutthroats in Henrys Lake and elsewhere are much safer than they used to be.

Idaho is more progressive than most states; still, it was only in 2001 that it fully implemented a policy of not stocking viable hatchery fish on top of wild populations.

Environmentalists are no more enlightened than sportsmen. In October 1997 the California Department of Fish and Game poisoned alien pike out of 4,000-acre Lake Davis in order to protect the endangered steelhead trout and chinook salmon of the Sacramento and San Joaquin river systems. The poison of choice—rotenone, the single most important tool of native-trout restorers—is derived from derris root. It is short-lived, applied at only 0.5 to 4 parts per million, and, after 69 years of use by fish managers, has not been seen to harm or even affect a human. Still, would-be protectors of water quality mounted vicious protests. They held all-night candlelight vigils, chained themselves to buoys, cursed, wept, and marched around the lake with placards that said things like "Burn in Hell, Fish & Game!" For crowd control the state deployed 270 uniformed officers, including a SWAT team. Now that pike are back in the lake, possibly because of sabotage, the state is too frightened to use rotenone again. Instead, it is proceeding with halfway measures, such as explosives, that can only suppress pike, not eliminate them.

Alpine lakes infested with hybrid cutthroats in Montana's Bob Marshall Wilderness are dribbling alien genes into pure westslope cutthroat populations in the Flathead drainage. To keep westslopes off the endangered-species list, the state's Fish, Wildlife and Parks department proposes to apply rotenone to a dozen of these lakes, then stock pure westslopes. But instead of rallying to the defense of this icon of American wilderness, the group Wilderness Watch is doing its best to kill the project, making ridiculous, untruthful pronouncements such as "Poison has no place in wilderness stewardship."

Managers create demand for hybrids just by supplying them. In Lake Superior, restoration of coasters—a race of giant brook trout—is finally getting under way. Ontario is doing great work. So are the Chippewa Indians. Minnesota is making a reasonable effort, but Michigan and Wisconsin are endangering the program by stocking Lake Superior with "splake," Frankenstein fish produced in hatcheries by crossing female lake trout with male brook trout. Not only do splake compete with coasters, the average angler can't tell them from coasters and winds up killing the latter. When I asked Wisconsin managers why they weren't doing more for coasters, I was told that the state has decided there's nothing special about them, that "a brook trout is a brook trout."

Such talk infuriates Robert Behnke of Colorado State University, the world's leading authority on trout and the man who rediscovered Lahontan and Bonneville cutthroats

after they'd been declared extinct. "A grape is also a grape," Behnke wrote me. "One species of grape (*Vitus vinifera*) is used in virtually all wine made in the world—reds, whites, best and worst. The grape-is-a-grape point of view is the most simplistic and would save money for wine drinkers, because the cheapest wines would be the same quality as the most expensive wines. I wouldn't want some of the managers [you] quote selecting wine for me or, for that matter, being in charge of fisheries programs where subtle genetic differences that may not show up in genetic analysis can be important."

I can't think of a finer rebuttal to the superstition that a "trout is a trout" than the southern Appalachian brook trout, which I first encountered in Great Smoky Mountains National Park. Fisheries biologist Matt Kulp had placed me in charge of measuring and releasing fish he'd just netted from a high-elevation rill, after briefly stunning them with an electric shock from a backpack generator. A big coaster weighs eight pounds, but the biggest Appalachian brook trout I handled that day—"huge" by park standards—weighed about four ounces. In sunlight, muted by the kind of cloud bank that gave these mountains their name, the belly of the little fish glowed campfire orange. The markings were different, too. Coasters and the brook trout of my Yankee woods have two or three rows of red spots along their chestnut flanks, but this one had seven. The dorsal fin was proportionately larger and marked with strange but lovely black stripes. Underfins, with the familiar ivory trim, seemed larger, too.

Rainbows, stocked by the park until 1976, have pushed the natives into the high country. So Kulp and his associates have been shocking the aliens and releasing them below natural barriers. Now that teams are working down into bigger water, shocking doesn't work. They need Antimycin, an incredibly selective and expensive fish poison that has a half-life of 40 minutes and is applied at 8 to 12 parts per billion. After rainbow advocates and chemophobes shrieked like scalded hogs, Kulp and his boss, Steve Moore, undertook exhaustive outreach. One sportsman, whom Moore thought he was literally going to have to fight, now admits to being "dead wrong," and, as with so much of the public, is utterly captivated by the South's native trout. So far the park has given 11.1 miles of stream back to the fish that belong here, and it's looking to restore an additional 40 miles, about all that's practical with current technology. Since the park has 750 miles of stream, there will be no shortage of rainbow fishing.

Managers have achieved another stunning success with the Gila trout of New Mexico, America's only endangered inland trout. I'd given up on the species when I inspected its habitat in 1994. In Black Canyon Creek, one of two perennial streams in the Aldo Leopold Wilderness, I encountered cattle in the water, knocking down the banks and defecating. So tolerant of cattle was the local Forest Service ranger that when I stopped in to see her I found cow pies on her office steps. When I asked Brub Stone, then a director of the Gila Fish and Gun Club, why his group opposed Gila trout restoration in Mineral and Willow creeks, he said: "They're using some kind of a fancy-name poison [Antimycin]. . . . Years ago they said the breast implant would not hurt women. My God, it's killing them, isn't it?"

In 1998, as the U.S. Fish and Wildlife Service and the state of New Mexico were preparing to reintroduce pure Gilas to Black Canyon Creek, they found rainbows, browns, and cutthroats (alien to the region and of undetermined race). The project had been sabotaged. Another stream, also cleansed of aliens, was apparently sabotaged. Grant County tried to kill the project—by preventing the use of Antimycin—with what it called the Pollution Nuisance Ordinance Act. Because of this ordinance (as well as the fact that managers wanted to avoid the hassle of evacuating, holding, then restocking rare dace and suckers) a 20-man crew equipped with backpack shockers removed 376 alien trout from Black Canyon Creek over the course of 88 days, a job that would have taken minutes with Antimycin. The reintroduced Gilas have been repro-

ducing for the past two years. The Forest Service has kicked out the cows, and streambanks are healing. In 1970 Gila trout survived in about 12 miles of stream in 4 drainages; now they inhabit about 80 miles in 13 drainages. "As far as we're concerned, we've satisfied the downlisting criteria," the Fish and Wildlife Service's Jim Brooks told me.

There's enough momentum in native-trout restoration that it might succeed nationwide if the environmental community gets behind it. The old-guard managers who flung trout around the country like Johnny Appleseed on applejack are dead, and, with only a few exceptions, their replacements are fiercely committed to natives. But most of these young scientists lack Moore's and Kulp's communication skills. For instance, they attempt to generate excitement for their work by pointing out that native trout are "indicator species," thereby implying that their worth is right up their with, say, a $200 water-sampling kit.

Managers need to quit trying to figure out what native trout can do for us and attempt a new approach. Maybe it starts with a simple statement that these fish are priceless works of art that need to be protected for themselves, for the species that need them, and for people who cherish them for what they are and because they are. ❖

Questions

1. How many of the cutthroat subspecies are extinct?
2. On Clear Creek, how much has the cutthroat spawning run declined?

3. What is Antimycin?

Answers are at the back of the book.

7

Wyoming's vast inland desert known as the Red Desert is being threatened by proposed oil and gas development schemes. Because the desert seems inhospitable on the surface, and because few people have heard of the Red Desert, critics argue that the Red Desert is not being treated like the national treasure that it is. An exploration of the diversity and importance of the Red Desert follows.

Hostile Beauty

Geoffrey O'Gara

National Wildlife, Washington, August/September 2002

AS THE LITTLE CESSNA DROPS ITS RIGHT WING and begins a slow turn east following the creekbed of the Dry Sandy, the Oregon Buttes loom before us like a blunt-faced whale surfacing from the sagebrush plains, barnacled with limber pine. In the passenger seat, conservationist Tom Bell tries to sort the snow-dusted trails, creeks and escarpments below us.

He is trying to pick out the Oregon Trail, one of several historic tracks across southwest Wyoming, where the mountains of the Continental Divide flatten out in a huge expanse of high, arid plain known as the Red Desert. Historic landmarks are only one of the unique features of the desert: Within an area of five million acres—bigger than the state of Connecticut—are huge traveling sand dunes, the largest migrating herd of pronghorns in North America, desert elk and the topographic wonder of the Great Divide Basin, a bowl from which no water escapes, east or west.

"Some people don't see the beauty, or the value," says Bell, who has spent a lifetime defending the Red Desert against those who would dig up its minerals, wipe out its wildlife or string power lines along the old wagon ruts. "But a vast inland desert at this high elevation is so unique."

There is no exact border to the Red Desert. At its heart is the Great Divide Basin, where the continent splits like a broken zipper just south of the Wind River Mountains and cups about 660,000 acres before coming back together south of Interstate 80 near Rawlins, Wyoming. Most descriptions of the Red Desert extend beyond the basin to include features such as the Killpecker Sand Dunes to the west and, to the south, Adobe Town, a wonderland of eroded sandstone pillars and cliffs near the Colorado border. What binds it all together is the unbroken openness, the lack of moisture and the lack of people.

Our pilot is taking us up because he's never had an opportunity to explore the Red Desert. He couldn't have found a better guide than Bell, a World War II bombardier who grew up just north of the desert, on a ranch outside of Lander. Bell is best known as the founder of the Colorado-based conservation publication *High Country News*. But Bell is also a wildlife biologist, and he knows where to look for desert elk: on the southern flank of the Oregon Buttes; in the Honeycomb Buttes, a maze of colorfully banded sandstone; and in the dunes, where snow refrigerated in sand melts to form wildlife-friendly ponds. He can also differentiate each scar on the landscape—even the tracks of seismic tests for oil and natural gas from the 1940s. "They claimed they could just cover that up," says the feisty 78-year-old, who recently was named the National Wildlife Federation's Conservationist of the Year. "That's a crock."

People like our pilot stay away because this country is, for all its stark beauty, wind-blown and inhospitable, and far from a coffee shop or phone booth. Few would guess, glancing out the window as they drive north from Rock Springs to Yellowstone, that they were looking at an ecological niche of immense importance.

"Species are contracting into this area from all over," said Erik Molvar, a biologist with Biodiversity Associates, a Wyoming conservation group that has studied the Red Desert for potential wilderness designation. "Take the mountain plover, one of the rare shorebirds that shows up here—it used to be a plains species, and this was the edge of its habitat. Now it's core habitat." Plovers, sage grouse and other species are declining to the east as the Great Plains are fenced and cultivated for agriculture; to the west, energy development and exotic species have invaded and compromised their habitat. The Red Desert may be their last refuge.

Molvar likes to point out lesser known but rare Red Desert residents such as the burrowing owl, blowout pen-stemon and the pygmy rabbit. But the high desert is also loaded with "celebrity" species, such as the golden eagles and ferruginous hawks that soar off the cliffs of Oregon Buttes, Steamboat Mountain and Continental Peak. Or the elk, which once migrated from the Yellowstone area, but now live year-round in the desert and number in the thousands.

Then there is the 40,000-strong pronghorn herd—you'll see them grazing on the hills, almost invisible against the tawny earth and snow, or racing like a flurry of leaves in the wind. Southwest Wyoming is their stronghold, and they journey annually from the Red Desert west and north along the Green River corridor, up into the Gros Ventre Mountains near Jackson Hole. The route is increasingly littered with oil and gas drilling rigs, "ranchette" subdivisions, and the roads that connect them.

Wyoming's reputation as the nation's "energy breadbasket"—with large reserves of coal, uranium, oil and gas—hovers over the Red Desert. A coal-bed methane drilling boom is underway in the Powder River Basin around Gillette, and industry sources say the methane gas deposits in the Green River Basin adjacent to the Red Desert are even larger. As a result, the energy industry is now turning its attention to the area over which we fly.

Much of the Red Desert is managed by the U.S. Bureau of Land Management (BLM), but it is broken up by sections of state lands and a checkerboard pattern of private and public ownership along the Interstate 80 corridor, which follows the old Union Pacific rail line. While BLM is looking at proposed wilderness areas in parts of the Red Desert, it has allowed drilling and exploration in many areas since the 1970s. In a region that gets less than ten inches of rain annually, these disturbances to the soil aren't easily erased.

As the plane follows the Dry Sandy east, it dawns on Bell that this is not the Oregon Trail immediately below. It is the bed of a narrow-gauge railway that once ran between an iron ore mine and mill near South Pass and the Union Pacific rail line in Rock Springs. Which reminds me, in the backseat, of another ride I'd taken a decade ago, clinging to the roof of a little diesel engine rolling down those tracks. On one side of us some pronghorn perked and dodged off into the hills; on the other, six feral horses raced. In the cab beneath me, John Mioczynski was at the controls. He had been hired by a salvage company when the mine shut down to patrol the tracks looking for vandalism.

Mioczynski is a wildlife biologist currently studying bighorn sheep for the Wyoming Game and Fish Department. He is one in a long line of desert rats who've fallen in love with a place which—with its hot dry summers and howling cold winters—shows no love for humans.

Mioczynski discovered it on a map: He was a college student in New York, looking for the most vacant place he could find to spend a summer. In 1967, he drove his jeep into the desert and lived on rabbits, sage grouse and wild plants while he learned the desert's flora and fauna. He found tipi rings, abandoned homesteads and parts of old wagons sticking up from quicksand. "I was all alone, and happy that way, but there were signs that people had been out here doing things," he says. "I began to see it was a pretty hospitable place. A hostile, beautiful, hospitable place."

Wandering around, he found an ancient awl made of yellow jasper, crocodile bones preserved in sediment and gastropod fossils. Fossils are everywhere in the exposed sandstones, telling the story of what was once a very different world. Some 50 million years ago, the Red Desert was part of a subtropical lake that covered much of southwestern Wyoming. The sands at the bottom of this lake, washed down from nearby mountains, today form the largest living dune system in the country, the Killpecker dunes.

The dunes stretch more than 50 miles west to east, from Boar's Tusk to Steamboat Mountain. Life on the dunes is mostly a night affair, when dune beetles, voles, shrews, white-footed mice and kangaroo rats are about, and the predators linked to them—bobcats, owls, golden eagles—follow. Snow collected on the lee side of dunes is covered and insulated by sand, banking ice for water throughout the year. The ponds here attract a surprising array of ducks and shorebirds, as well as desert elk.

These elk are not the same herd that was noted by hunters back in the nineteenth century. Settlement and hunting wiped out most of those animals by the 1940s, when the Wyoming Game and Fish Department transplanted 86 elk from the Jackson area to the vicinity of Boar's Tusk and the Killpecker dunes. Wildlife managers hoped the elk would migrate back and forth from the Jackson area, as they had historically, which would reduce the necessity for feeding them through the winter at the National Elk Refuge.

But the transplanted elk realized, like Mioczynski, that there was a lot of nourishment hidden beneath the desert's stern face, especially around Steamboat Mountain, and there they stay year-round. The elk population has grown far beyond the herd of 500 envisioned, and another herd of elk now comes down from the Wind River Mountains to winter along the Sweetwater River, on the north edge of the desert.

Leonard Hay, a banker and rancher from Rock Springs, was there when the elk were reintroduced, and like many locals involved in grazing and energy development in the Red Desert, he scoffs at conservationists' efforts to protect the area. "I enjoy drilling on my land," says the outspoken Hay, now in his 80s, "and those rig workers usually have a rifle in the pickup, which means less coyotes."

Hay spoke to me during a meeting held by BLM in Rock Springs last January. The meeting was one of several to consider management options for a 622,000-acre parcel of the Red Desert called the Jack Morrow Hills, which includes Steamboat Mountain, the Oregon Buttes, the sand dunes and the Honeycomb Buttes. Milder-mannered ranchers than Hay expressed fear that their grazing rights on public lands might be endangered, but many conservationists are willing to work with ranchers—their fears for the Red Desert focus more on the energy industry.

BLM's proposal to allow more oil and gas development in the Jack Morrow Hills—where 150 wells have been sunk in the past—has thrust the Red Desert and the local BLM office into a rather unfamiliar national spotlight. "We have no idea what the administration is going to do with Jack Morrow Hills, but public comment favors a conservation alternative," says Mac Blewer, an organizer for the Wyoming Outdoor Council. "It's a treasure and deserves the highest protection."

The effort to protect the Red Desert goes back more than a century. In 1898, sportsmen proposed a federal "winter game reserve" in the desert. In 1935, Wyoming Governor Leslie Miler proposed a Great Divide Basin National Park, citing the rich history of the area. Wyoming geologist Dave Love and Tom Bell each met with Interior Secretary Stewart Udall in the 1960s to lobby for protection at some level, whether it be a park, an antelope preserve or a national landmark.

Love acknowledges that there is not just oil and gas in the Red Desert, there are coal seams "50 feet thick, more than anyone knows." There are open-pit coal mines in the checkerboard of private lands between Rock Springs and Rawlins, and the remnants of an abandoned uranium mine and mill.

Add to that the damage left by over-grazing generations ago, and it's clear that this is not an area with a pristine past—a point which detractors are quick to make. But its vistas are still open, and its history goes much further back than mere centuries. As we fly north, Bell describes the Red Desert as "one of our very best wildlands, and one of our last," and points to places below where one might find dinosaur remains or ancient sharks' teeth.

That's the sort of thing John Mioczynski might have picked up during his early wanderings in the desert. He used to put the things he found in cigar boxes—shards of quartzite, bones, arrowheads, fossils. But one day he took his collection back out in the desert, to the areas where he'd found the artifacts, and left them there.

"After that, I'd pick up an antelope skull, walk with it for awhile, then put it down," says Mioczynski. "Everyone's out there taking things. You don't take things from a place that's sacred." ❖

Questions

1. How big is the Red Desert?
2. Why does Wyoming have a reputation as the nation's "energy breadbasket"?
3. How much rain does the Red Desert receive?

Answers are at the back of the book.

8

Linking our wildlife preserves with "wildlife corridors" may be a viable way to preserve wildlife diversity and prosperity. According to Wilson and MacArthur's "Theory of Island Biogeography," isolation of animal habitats can lead to a loss in biodiversity. North America's wildlife preserves are effectively land-locked "islands." Wildlife corridors provide migratory routes and allow for future evolution of wildlife. While there is some discussion over whether or not corridors will be sufficiently useful or effective, efforts to begin and expand corridor projects are already underway.

Wilding America

Elizabeth Royte

Discover, September 2002

THE MOUNTAIN LION WAS HEALTHY, MALE, AND YOUNG. He was born in the Santa Ana Mountains of southern California, probably in the dry, rugged hills near the seaside town of San Clemente. As a juvenile, he wandered through chaparral, hunting mule deer, jackrabbit, bobcat, and coyote. At 18 months, the lion—known as M6 to the scientist who tracked his movements—began to roam farther, looking for a home range of his own and a mate.

One night M6 headed north. At midday he rested; when darkness fell, he resumed his trek. About 50 miles into his journey, he left the conifers of the higher peaks in the Cleveland National Forest and dropped down into the sage scrub of Coal Canyon. Its stony creek bed led him into a broad, sandy outwash. Here M6 took stock of his predicament. An eight-lane freeway, Highway 91, the major thoroughfare from Riverside County to Los Angeles, blocked his progress. Hundreds of cars every hour streamed past. M6 sniffed out a derelict underpass. It was noisy and uninviting, but he made it through, leaving the highway behind and entering the relative calm of Chino Hills State Park. For 187 days M6 stayed put, patrolling 12,000 acres of low, grassy hills. Then he started to move again. Chino Hills, apparently, wasn't big enough.

Twenty-two times over the next 19 months, M6 made the journey back and forth under Highway 91. He became a street-smart lion, but the passage was always perilous. To reach the canyon, M6 had to work his way across two shrubless golf courses, which offered little in the way of protective cover, and past a stable. Before he could get to the freeway, he had to cross a double set of busy railroad tracks. This was, by any sentient being's measure, difficult terrain. Arc lights glared; traffic roared.

Despite the obstacles, M6 stitched together an hourglass-shaped home territory of some 168 square miles between the Chino Hills and the Santa Ana Mountains. Coal Canyon connected the lobes of the hourglass. For M6 it had become a corridor of life and death. The lion could easily have been hit by a car (at least six mountain lions were killed by cars in southern California last year), kicked by a horse, or flattened by Amtrak. But M6 had little choice: The Chino Hills contained only enough prey to support one or two female lions. If M6 wanted to pass on his genes, he had to survive the Coal Canyon choke point.

And life for this lion was about to get even harder. A developer had plans for 652 acres just south of the freeway: 1,500 houses, plus all the usual gas stations and fast-food outlets that attend the birth of a neighborhood. Building and paving would sever the already tenuous connection to Coal Canyon. According to Paul Beier, the scientist who collared M6 and tracked him for months, "the loss of this corridor would guarantee the extinction of the mountain lion from Chino Hills and endanger the entire population of lions in the Santa Anas."

Wildlife corridors have a wonderful, utilitarian simplicity, especially in a crowded place like southern California, where the prospect of creating large new preserves is relatively limited. Here many biologists believe it makes sense to connect smaller, established parks—islands of biological diversity—with wildlife corridors. These may be blocks of ranch land, ribbons of land alongside streams, or highway underpasses. Strung together like green beads on a necklace, each piece of land could become part of a larger whole. Animals like M6 would be able to move freely, get enough food, woo a mate, and reproduce. "We can tweak the margins of our parks and wilderness areas, extend them a little here or there," says Beier. "But to make them viable over the long term, we have to think about how we're going to link them."

Linkages in southern California tend to be modest in size, but some conservationists, thinking more ambitiously, envision linkages as an integral part of a far grander scheme: the re-wilding of the entire continent. Across North America, large parklands could be connected either through the acquisition of additional land parcels, which would form corridors, or by retrofitting roadways with underpasses that let animals move freely between reserves. Populations of creatures that have been driven from their native habitats—wolves, for example, or black-footed ferrets—would be restored. The time for such conservation megaprojects, writes the evolutionary biologist Edward O. Wilson in *The Future of Life,* "is now, because the windows of opportunity are closing fast."

The contemplation of wildlife corridors grew out of *The Theory of Island Biogeography,* written by Wilson and the ecologist Robert MacArthur in 1967. The theory states, in simplest terms, that bigger islands closer to mainlands have more species than islands that are smaller and more isolated. Several decades ago biologists realized that the same theory could be applied to protected parks that had become isolated as developments and roads tightened around them. Fragmenting forces—whether a golf course, a clear-cut, or a four-lane highway—effectively make islands out of nature reserves. Even relatively large parcels of land can be doomed to islandhood, according to one often-cited study. Bryce Canyon National Park, Lassen Volcanic National Park, and Zion National Park have each lost about 40 percent of their larger mammal species since they were founded. Humans either killed them off directly or reduced their habitats.

Population biology theory suggests that without migratory routes, animals in small parks, like animals on small islands, may be subject to the same inbreeding pressures as zoo populations. Corridors may counter this effect by allowing a dwindling population in one area to be supplemented by individuals from another. They allow adolescents to disperse and genes to flow among populations. Animals can migrate to establish new home ranges, as M6 did, and follow their prey from higher to lower elevations as the seasons change. They can move in response to short-term environmental change such as fire or drought and long-term environmental change such as rising temperatures. "Protecting linkages will ensure these species don't blink out in the short term," says Kristeen Penrod, executive director of the South Coast Wildlands Project. "In the long term, they will let these species evolve."

One sunny weekend, Beier, Penrod, and about 200 biologists, land managers, and town planners met at the San Diego Zoo to plot the future of California's natural heritage. Poring over maps and animal censuses, they identified 232 missing linkages deemed critical for preserving the state's biodiversity. Out of that total, 60 potential linkages lay in the south coastal region of southern California. This richness of local opportunity surprised no one. Thanks to its Mediterranean climate and its mountainous terrain, the area between Los Angeles and San Diego contains 2,500 plant species that live nowhere else in the world. A lot of plants at the bottom of the food chain mean a lot of animal diversity higher up. The area is, in conservationists' jargon, a biological hot spot.

Unfortunately, a great deal of that biodiversity is about to disappear. Southern California has more threatened and endangered species than any other region in the continental United States. By some estimates, 200 plant species and 200 animal species—from bighorn sheep to foxes and butterflies—are imperiled.

The problem is too many people. Supermarkets, condos, and offices have already wiped out about 90 percent of the region's wetlands. This constitutes a significant environmental blow because wetlands filter pollution, absorb storm runoff, and provide habitat for thousands of plant and animal species. Furthermore, most of San Diego's 2.8 million residents live within several miles of the Pacific Ocean, where nearly all the coastal sage scrub has already been plowed under to build houses and shopping centers. Coastal sage scrub alone contains more than 35 plant species, two insects, seven reptiles, four birds, and seven mammals that are listed as endangered or are candidates to be listed.

To document which creatures are hanging on in the remaining green pockets—and how they eat, breed, migrate, and die—biologists are doing some odd things. In the chaparral, they're dripping the anal scent of bobcats onto rocks to lure felines toward camera traps. They must prove how many bobcats are present and whether they're successfully breeding to make a case for preserving bobcat habitat. In the mountains, biologists are setting foothold snares to capture lions, which they'll fit with radio collars. They're dusting animal paths with gypsum powder so they can quantify the footprints of anything that walks by.

Mountain lions get a lot of attention from conservation biologists. They need large home ranges and large populations of animals, like deer, to eat. Youngsters need plenty of room to disperse and mate. If lions, which have stalked these mountains for millennia and shaped the evolution of numerous interconnected species, don't get more space soon, says Beier, they will simply disappear. Forever.

The extinction of mountain lions might please homeowners with pet Pomeranians and calicos that wander near the Chino Hills, but should the species disappear locally, a cascade of effects would ripple through the food chain. Deer would proliferate, overbrowsing forests and shrubs that harbor smaller animals. Free of their feline nemesis, populations of middle-size predators, such as skunks, raccoons, gray foxes, and opossums, would boom, decimating populations of smaller prey animals. In addition to mice and voles, nesting birds would take a hit, including the endangered California Gnatcatcher. Birds keep invertebrate populations in check, and they also move around a lot of seeds, which grow into plants that feed other birds, butterflies, and other small animals.

About a century ago, mountain lions could be found in nearly every continental state. Now the only hope of seeing them again lies in the work of conservationists who design networks to restore the wild cats' movement. In the Southeast, for example, connected parklands could enable a young male panther in the Florida Everglades to stake out a home territory near his relatives in Georgia's Appalachians.

The idea of restoring the movement of large carnivores across America, of establishing wilderness corridors that stretch across the continent, seemed far-fetched when an environmental group called the Wildlands Project proposed it more than 10 years ago. But the vision has worked its way into the mainstream. The group's goals, writes Paul Ehrlich,

professor of population studies at Stanford University's Center for Conservation Biology, "have now been embraced broadly as the only realistic strategy for ending the extinction crisis."

In the West, biologists have already drawn a blueprint for connecting the upland habitats of southern New Mexico, southern Arizona, and northern Mexico. The plan calls for the recovery of all large carnivores native to the region, for the restoration of watersheds, forests, and natural fire regimes, the establishment of movement corridors, and the control of non-native species. Similar goals are being set for a 2,000-mile swath of forest that stretches from the Yukon all the way to Yellowstone National Park, and for the southern Rockies, from Wyoming to the Sierra Madre in Mexico. In the East, conservationists envision an Appalachian corridor of more or less continuous forest from western Pennsylvania into eastern Kentucky.

"It's an incredible dream," says Michael Soulž, emeritus professor of environmental studies at the University of California at Santa Cruz and a director at the Wildlands Project. He imagines jaguars, ocelots, and jaguarundis prowling over their historical range in Texas, southern New Mexico, and Arizona, as well as wolves running "through most of the mountains. There's plenty of habitat and food for these animals." Grizzlies could range in a nearly continuous chain from the Sonoran mountains of Mexico to the Yukon.

The Wildlands Project doesn't call for dismantling roads, pipelines, or shopping centers. Instead, members hope to expand, connect, and restore wilderness areas by consolidating new developments and adapting buildings and other structures already in place. In the northern Rockies, for example, the group would like to limit construction of shopping centers along a major highway that threatens to sever the flow of large carnivores between Canada and the United States. The plan calls for retrofitting the highway with underpasses for wild animals.

Linkages in southern California don't have the grandeur of a Yukon-to-Yellowstone wildlife corridor. But some local conservationists believe southern California can be a model for planning across the country. "Other National Park Service people look here for lessons," says Ray Sauvajot of the Santa Monica Mountains National Recreation Area. "All our parks are being encroached upon. Even the Yosemite Valley floor is becoming fragmented." Southern California is unique, both for its intense population pressure and for how much biodiversity is at stake. "But if corridors can work here," Sauvajot says, "then they'll work anywhere."

Although Sauvajot is talking about political success, some scientists question the very assumptions of wildlife corridors. They ask if corridors might usher disease or exotic species from one reserve to another. So far, there's no evidence that this has occurred. But it's difficult to design a rigorous study of corridor utility. For comparison's sake, researchers would have to build a corridor and also remove a corridor, then track animal movement and reproductive success for years. Biologists have a great deal of empirical evidence that animals use corridors, but they don't know yet if the right animals are using them at the right time, or if some predators might use them as avenues to kill prey that they wouldn't normally encounter.

"In most cases, there is no direct evidence that corridors are used for movement or that movement is important to the persistence of a population," says biologist Dan Simberloff of the University of Tennessee at Knoxville. "There has been no real study of their cost-effectiveness." He believes money spent on corridors could be better spent buying larger parcels of land. In southern California, the debate seems irrelevant. There aren't many large parcels left to buy.

Beier says little is known about how well corridors function. In 1998 he and biologist Reed Noss of Oregon State University reviewed 32 studies of wildlife corridors and found that fewer than half provided persuasive data. Still, they concluded, "well-designed studies suggest that corridors are valuable conservation tools." Beier invokes the precautionary principle: In the face of serious threat, a lack of certainty shouldn't prevent taking action to prevent or minimize that threat. "It may be better to build a corridor and find out what happens than to lose the land to development. We don't want to find out what will happen if we lose these corridors."

Taking a contrarian view, Simberloff argues for "management procedures over buying corridors. We could restore habitat, move animals to other reserves, hire more wardens, and build fences to keep lions off highways."

While biologists battle, private conservancies and California's Department of Fish and Game are busy buying land for corridors. They either buy outright from willing sellers, or they broker trades, wherein an owner who holds undeveloped land deemed critical to a corridor exchanges it for state-owned land of equal value that isn't. Land may also become part of a corridor through conservation easement: The Nature Conservancy, for example, buys land, then resells parcels with restricted development rights.

Pat and Jeanine Stambersky bought five sage-dotted acres smack in the middle of the Tenaja Corridor, a three-mile-long swath that lies just west of Riverside County's teeming exurban developments. The corridor connects the 8,300-acre Santa Rosa Ecological Reserve with a 160,000-acre section of the Cleveland National Forest. In theory, a lion could travel south from Chino Hills State Park through the national forest and all the way to the Santa Rosa reserve, about 70 miles away.

While some of their neighbors have easements, the Stamberskys have an understanding with the Nature Conservancy that limits building on the land, stipulates that any fences must be friendly to wildlife, and stops them from clearing native vegetation. "We didn't think we were giving up anything," says Jeanine Stambersky. "We moved out here to be with nature." Every morning she watches a pack of coyotes trot past her house.

Judy Kollar bought land in the corridor because "I knew this would be a model for how people could live in semi-wilderness areas and make as little impact as possible. I'd consider the plan a failure if I didn't see coyotes up here." A fifth-generation southern Californian, Kollar takes natural heritage personally. "I'm a connoisseur of the local landscape," she says. "I have an aesthetic reaction to its being plowed under."

There are others like Kollar. In Orange County, Claire Schlotterbeck rallied support for mountain lions and other species that needed to pass through the bottleneck of Coal Canyon. After a two-year effort, her advocacy group, Hills for Everyone, persuaded the state to put up $14.7 million to help buy out the real estate developer who had planned on building condos south of Highway 91 and to purchase an additional 32 acres of land just to the north.

"The department of transportation is restoring the underpass," says Geary Hund, a state parks department wildlife ecologist who worked with Schlotterbeck. "We'll get rid of the lights and put up some sound barriers. We'll pull up the pavement under the freeway and set up some fencing to steer animals off the highway and down under." Native vegetation will be planted on the compacted slopes; Coal Creek's concrete waterway will be ripped up and half its natural flow restored. Instead of cement, he says, imagine willows and mule fat shrubs. Then imagine birdsong in the air, and butterflies.

And imagine M6 still prowling around. When the lion study was over, Beier removed the radio collar. He likes to imagine that M6 is going strong. "He's the only male in that area," Beier says. "He's pretty tough. I bet he found a female." He pauses to think, then adds, "I sure hope so." ❖

Questions

1. Why are wildlife corridors ideal for a crowded place like southern California?
2. The contemplation of wildlife corridors was provoked by what?
3. What odd things are biologists doing to document which creatures are hanging on in the remaining green pockets and record how they eat, breed, migrate, and die?

Answers are at the back of the book.

Section Three

Resource Use and Management

9

Although problems with energy resource scarcity are surfacing, few countries are making a valiant effort to change their energy mix. However, of the small number of countries that are making the switch to renewable energy sources their results are promising. Countries that led the way in renewable energy use include Denmark, Germany, and Spain. Overall, Europe's reliance on wind energy now constitutes 70% of wind power used continent wide. Incentive programs, such as tax credits and subsides are garnishing the way wind power looks on the global energy scale. Even a couple of states in the Pacific Northwest are finding wind power to be not only an awesome sight but also a competitive source of energy.

The Winds of Change

Margot Roosevelt

Time, August 26, 2002 (Special Report: Green Century)

OVER THE COLUMBIA RIVER, on a high desert ridge, the world's largest wind farm sprawls across 50 sq. mi. of Oregon and Washington. When the last of its 460 turbines are installed, this postmodern power plant will offer clean electricity to 70,000 homes and businesses. Every month hundreds of tourists come to gawk at its fiber-glass blades, twirling with balletic grace atop 160-ft. poles. "People are in awe of wind power," says Anne Walsh, community-relations manager of the Stateline Energy Center.

And guess what? Wind is becoming more than a quixotic sideshow. It's now the world's fastest growing power source—a high-tech challenge to the coal mines, oil rigs, nuclear reactors and hydroelectric dams that seem, well, so 20th century. Experts say wind could provide up to 12% of the earth's electricity within two decades. Wind farms in Texas, Oregon, Kansas and elsewhere helped lift U.S. wind-energy output 66% last year, and an additional $3 billion in American projects are in the works. "Wind is competitive," wrote Mark Moody-Stewart, the former chairman of Royal Dutch/Shell who now co-chairs an alternative-energy task force for the Group of Eight, in a recent report. "This is not something to look forward to for the future—it is here today."

The promise is tantalizing. Windmills generate renewable power, so called because the source of the energy, wind, is continually renewed by nature (ditto for solar cells, which are powered by the sun; geothermal systems, which use the earth's heat; and hydroelectricity, which flows from dams). Unlike oil and coal deposits, renewable energy can't be exhausted, at least not until the sun burns out billions of years from now and earth goes cold.

Skeptics may recall the burst of enthusiasm for conservation and renewable power when oil prices quadrupled in the 1970s. State-funded energy research and development surged, while tax incentives boosted solar, wind and other alternatives to petroleum and the atom. But once oil supplies loosened and prices dropped, governments lost interest. In the U.S., rules requiring more fuel-efficient cars were rolled back. In California, subsidies evaporated, pushing wind companies into bankruptcy. "It is a moral disgrace that we have done so little to reduce our dependence on imported oil and oil generally," says Reid Detchon, a former U.S. Energy Department official who now consults for the United Nations Foundation.

But the need to diversify is now more urgent and the consensus to do so is greater than when OPEC first played bully. Global energy demand is expected to triple by mid-century. The earth is unlikely to run out of fossil fuels by then, given its vast reserves of coal, but it seems unthinkable that we will continue to use them as we do now, for nearly 80% of our energy. It's not just a question of supply and price, or even of the diseases caused by filthy air. We know that global warming from heat-trapping carbon dioxide, a by-product of fossil-fuel burning, threatens to cause chaos with the world's climate. And the terrorist assault on the World Trade Center raises other scary scenarios: how much easier would it be to crack open the Trans-Alaska pipeline and how much deadlier would it be to bomb a nuclear plant than to attack a wind farm?

Clean energy has a long way to go. Only 2.2% of the world's energy comes from "new" renewables such as small hydroelectric dams, wind, solar and geothermal. (Traditional renewable energy from large dams provides another 2.2%.) How to boost that share—and at what pace—is debated in industrialized nations—from Japan, which imports 99.7% of its oil, to Germany, where the nearby Chernobyl accident turned the public against nuclear plants, to the U.S., where the Bush Administration has strong ties to the oil industry. But the momentum toward clean renewables is undeniable. Globally, solar- and wind-energy output is growing more than 30% annually—far faster than conventional fuels—and their cost is plummeting. "We are on the cusp of an energy revolution," says Christopher Flavin, president of the Worldwatch Institute, a Washington nonprofit. "It will be as profound as the one that ushered in the age of oil a century ago."

Even oil companies are trying to cash in on the decarbonization trend. The world has gradually moved toward cleaner fuels—from wood to coal, from coal to oil and from oil to natural gas. Renewables are the next step. Royal Dutch/Shell has pledged to spend up to $1 billion on renewables through the next five years. Japanese manufacturers, led by Sharp and Kyocera, have moved aggressively into photovoltaic cells, which turn sunlight into electricity. And in April General Electric snapped up Enron Wind from the bankrupt energy giant. "We are on a journey to a lower-carbon world," says Graham Baxter, an executive at Britain's BP, which is building a $100 million solar plant in Spain.

How soon we reach an era of clean, inexhaustible energy depends on technology. Solar and wind energies are intermittent: when the sky is cloudy or the breeze dies down, fossil fuel or nuclear plants must kick in to compensate. But scientists are working on better ways to store electricity from renewable sources. Current from wind, solar or geothermal energy can be used to extract hydrogen from water molecules. In the future, hydrogen could be stored in tanks, and when energy is needed, the gas could be run through a fuel cell, a device that combines hydrogen with oxygen. The result: pollution-free electricity, with water as the only by-product. Already fuel-cell buses, cars and small generators are being tested. Eventually, some visionaries say, fuel cells placed in individual buildings could replace many of today's giant electric plants. But that will not happen unless the technology is refined and the cost drops. "A hydrogen economy," says Alan Nogee of the Union of Concerned Scientists, a U.S. environmental group, "is the Holy Grail."

While the developed nations debate how to fuel their power plants, however, some 1.6 billion people—a quarter of the globe's population—have no access to electricity or gasoline. They cannot refrigerate food or medicine, pump well water, power a tractor, make a phone call or turn on an electric light to do homework. Many spend their days collecting firewood and cow dung, burning it in primitive stoves that belch smoke into their lungs. To emerge from poverty, they need modern energy. And renewables can help, from village-scale hydro power to household photovoltaic systems to bio-gas stoves that convert dung into fuel. More than a million rural homes in developing countries get electricity from solar cells. "The potential is enormous," says Anil Cabraal, an energy specialist for the World Bank, which has helped finance 500,000 residential solar systems from Argentina to Sri Lanka.

Ultimately, the earth can meet its energy needs without fouling the environment. "But it won't happen," asserts Thomas Johansson, an energy adviser to the United Nations Development Program, "without the political will." To begin with, widespread government subsidies for fossil fuels and nuclear energy—estimated at some $150 billion per year—must be dismantled to level the playing field for renewables. Policymakers must factor in the price of pollution: coal plants are more expensive than renewable power when one includes the cost of scrubbers on smokestacks and the expense of health care for coal-related illnesses; nuclear energy costs would soar without government insurance. Environmentalists are calling for taxes on carbon to slow the growth of fossil-fuel use.

Another way to increase renewables' share of the energy mix is to reduce the use of conventional fuel through efficiency incentives. Experts estimate that efficiency could slash the globe's projected energy consumption by a third. Strict standards can cut energy use in everything from air conditioners to cars. Compact fluorescent lamps use a quarter of the electricity of incandescent bulbs to provide the same amount of light.

Governments are increasingly forcing utilities to make it easier for windmill and solar-panel owners to connect to the grid and get credit for providing extra electricity they don't use. Governments are also pressuring utilities to meet targets for renewable sources of energy. The European Union, for instance, is requiring its members to boost electricity from renewables to 22% of production within the next eight years. Brazil plans to push a global standard at the World Summit on Sustainable Development in Johannesburg this month.

On the road to enlightened energy policy, a few countries offer models of reform. More than a decade ago, Denmark required utilities to purchase any available renewable energy and pay a premium price; today the country gets 18% of its electricity from wind. Thanks largely to Germany and Spain, which have enacted vigorous incentives for renewables, Europe today accounts for 70% of the world's wind power. In Japan 80,000 households have installed solar roof panels since the government offered generous subsidies in 1994; consequently, Japan has displaced the U.S. as the world's leading manufacturer of photovoltaics. India established a fund that has lent $1.1 billion to alternative-energy projects; the country is now the globe's fifth largest generator of wind and solar power. Iceland, which lies on a hotbed of underground volcanic activity, uses that geothermal energy to heat 90% of its buildings. The island nation is planning to use geothermal and hydroelectric power to produce large amounts of hydrogen, creating the world's first hydrogen economy.

Such examples show that the future "is more a matter of choice than destiny," as Brazilian physicist Jose Goldemberg, the chairman of a recent United Nations energy study, put it. On the windy border of Washington and Oregon, citizen groups are already making a choice. They have pressured utilities to invest in green energy, and a federal tax credit has made it more profitable. "It's the right thing to do," says Vito Giarrusso, manager of the Stateline wind farm, "to help our little piece of the earth." ❖

Questions

1. How much of the earth's electricity could wind provide within two decades?
2. How much of the globe's population has no access to electricity or gasoline?
3. What country uses geothermal energy to heat 90% of its buildings?

Answers are at the back of the book.

10

As alarming statistics on climate change are just beginning to seep into the world psyche, scientists have demanded that the use of clean energy sources be maximized for future decades. Fossil fuels such as coal and oil have created a deadly reliance that not only pollutes the land, air, and sea but also accelerates climate change at a staggering rate. Europe and Japan have agreed to do their part to meet deadlines to cut greenhouse gas emissions by signing the Kyoto Protocol, while the United States, the nation that produces the most emissions, has yet to sign.

Scientists Say a Quest for Clean Energy Must Begin Now

Andrew C. Revkin

New York Times, November 1, 2002

To SUPPLY ENERGY NEEDS 50 YEARS from now without further influencing the climate, up to three times the total amount of energy now generated using coal, oil, and other fossil fuels will have to be produced using methods that generate no heat-trapping greenhouse gases, the scientists said in today's issue of the journal Science. In addition, they said, the use of fossil fuels will have to decline, and to achieve these goals research will have to begin immediately.

Without prompt action, the atmosphere's concentration of greenhouse gases, mainly carbon dioxide from burning fossil fuels, is expected to double from pre-industrial levels by the end of this century, the scientists said.

"A broad range of intensive research and development is urgently needed to produce technological options that can allow both climate stabilization and economic development," the team said.

The researchers called for intensive new efforts to improve existing technologies and develop others like fusion reactors or space-based solar power plants. They did not estimate how much such a research effort would cost, but it is considered likely to run into tens of billions of dollars in government and private funds.

The researchers, a team of 18 scientists from an array of academic, federal, and private research centers, said many options should be explored because some were bound to fail and success, somewhere, was essential.

The researchers all work at institutions that might themselves benefit from increased energy research spending, but other experts not involved in the work said the new analysis was an important, and sobering, refinement of earlier projections.

As they now exist, most energy technologies, the scientists said, "have severe deficiencies." Solar panels, new nuclear power options, windmills, filters for fossil fuel emissions and other options are either inadequate or require vastly more research and development than is currently planned in the United States or elsewhere, they said.

The assessment contrasts with an analysis of climate-friendly energy options made last year by the Intergovernmental Panel on Climate Change, an international panel of experts that works under United Nations auspices. That analysis concluded that existing technologies, diligently applied, would solve much of the problem.

One author of the new analysis, Dr. Haroon S. Kheshgi, is a chemical engineer for Exxon Mobil, whose primary focus remains oil, which along with coal generates most of the carbon dioxide accumulating in the air from human activities.

Still, Dr. Kheshgi said on Thursday that "climate change is a serious risk" requiring a shift away from fossil fuels. "You need a quantum jump in technology," he said. "What we're talking about here is a 50- to 100-year time scale."

Dr. Martin I. Hoffert, the lead author and a New York University physics professor, said he was convinced the technological hurdles could be overcome, but worried that the public and elected officials may not see the urgency.

In interviews, several of the authors and other experts said there were few signs that major industrial nations were ready to engage in an ambitious quest for clean energy.

Prof. Richard L. Schmalensee, a climate-policy expert and the dean of the Massachusetts Institute of Technology Sloan School of Management, said the issue of climate change remained too complex and contentious to generate the requisite focus. "There is no substitute for political will," he said.

The Bush administration has resisted sharp shifts in energy policy while Europe and Japan have accepted a climate treaty, the Kyoto Protocol, that includes binding deadlines for modest cuts in gas emissions. At international climate talks that end today in New Delhi, leaders of developing countries rejected limits on their fast-growing use of fossil fuels, saying rich countries should act first.

President Bush has called for more research, led by the Energy Department, on many of the technologies examined in the new analysis. But some energy and climate experts said the extent of the challenge would likely require far more focus and money than now exists.

Among the possibilities are space-based arrays of solar panels that might beam energy to earth using microwaves. The panel described various nuclear options, including the still-distant fusion option and new designs for fission-based power plants that might overcome limits on uranium and other fuels.

Planting forests, which absorb carbon dioxide, cannot possibly keep up with the anticipated growth in energy use as developing countries become industrialized and as global population rises toward nine billion or more, the panel said.

Some environmental campaigners criticized the study's focus on still-distant technologies, saying it could distract from the need to do what is possible now to reduce emissions of warming gases.

"Techno-fixes are pipe dreams in many cases," said Kert Davies, research director for Greenpeace, which has been conducting a broad campaign against Exxon Mobil. "The real solution," he said, "is cutting the use of fossil fuels by any means necessary." ❖

Questions

1. Without prompt action, the atmosphere's concentration of greenhouse gases is expected to increase by how much?
2. What countries have accepted the Kyoto Protocol and what does the climate treaty entail?
3. According to Kert Davies, research director for Greenpeace, what is the real solution to the global warming problem?

Answers are at the back of the book.

11

A new report has established a link between scarcity of water resources and poverty. This report, called the *Water Poverty Index*, was created by researchers from the Centre for Ecology and Hydrology in Britain and from the World Water Council. The report compares five measures of a country's water situation to determine how to best address the water concerns of different countries. Water capacity, resources, access, use, and environmental considerations are all weighed when an assessment of a country's water situation is done. Soon, wealthy nations will need to produce food for other nations instead of importing it in order to maintain an international water balance.

Link Seen Between Water Scarcity and Poverty

Sanjay Suri

Global Information Network, New York, December 12, 2002

A NEW REPORT BY THE WATER POVERTY INDEX (WPI) establishes a clear link between water scarcity and poverty, a top expert says.

William Cosgrove, vice president of the World Water Council, told IPS that wealthy nations will soon have to start producing food for other nations rather than buying it from them in order to preserve a water balance, in buying food from other places, countries such as the U.S. and Canada "are really purchasing a water resource."

Export of food grown often with use of scarce water means "a lot of water is being moved around the planet," Cosgrove said. "These wealthy nations will soon be called upon to produce more food with their water resources for other countries, so they will have to learn to use water better."

"That is what our instincts have always told us, but this is the first time that the link has been established and analyzed," Cosgrove said.

The Water Poverty Index was developed by a team of 31 researchers in consultation with more than 100 water professionals from around the world. The researchers were from the Centre for Ecology and Hydrology in Britain and from the World Water Council.

The top 10 water-rich nations in the world are: Finland, Canada, Iceland, Norway, Guyana, Surinam, Austria, Ireland, Sweden and Switzerland. The 10 countries lowest on the WPI are all in the developing world—Haiti (the lowest), Niger, Ethiopia, Eritrea, Malawi, Djibouti, Chad, Benin, Rwanda, and Burundi.

"The links between poverty, social deprivation, environmental integrity, water availability and health becomes clearer in the WPI, enabling policy makers and stakeholders to identify where problems exist and the appropriate measures to deal with their causes," says Caroline Sullivan, who led an interdisciplinary team to develop the WPI concept at the Centre for Ecology & Hydrology.

The link between poverty and water shortage established by the report will be presented as a prime subject at the Third World Water Forum in Kyoto in March. About 10,000 government officials, representatives of international and nongovernmental organizations, industry and water experts will discuss the world water crisis and its solutions at the forum.

"The international Water Poverty Index demonstrates that it is not the amount of water resources available that determine poverty levels in a country, but the effectiveness of how you use those resources," says Dr. Sullivan. Haiti and the Dominican Republic share the Caribbean island of Hispaniola and have more or less the same amount of water, but Haiti ranks last at 147th while the Dominican Republic ranks 64th.

The report establishes that some of the world's richest nations such as the United States and Japan fare poorly in water ranking, while some developing countries score in the top ten. In order to show where the best and worst water situations exist, the report grades 147 countries according to five measures—resources, access, capacity, use and environmental impact.

Following is a global snapshot on each of the five counts:

Capacity: This defines the ability to purchase, manage and lobby for improved water, education and health. The top five here are Iceland, Ireland, Spain, Japan and Austria. These countries have high incomes and healthy and well-educated populations. The bottom five are Sierra Leone, Niger, Guinea-Bissau, Mali and the Central African Republic. Besides being among the world's poorest, these countries also suffer from inadequate health and education provision.

Resources: These measure the per-capita volume of surface and groundwater resources that can be drawn upon. The top five are Iceland, Surinam, Guyana, Congo and Papua New Guinea. The bottom five are United Arab Emirates, Kuwait, Saudi Arabia, Jordan and Israel. The top countries have abundant resources, and small populations in relation to the resources. The bottom countries are all in desert areas but despite the scarcity of water, Israel, Kuwait and Saudi Arabia are in the top 50 percent as measured by the WPI, reflecting their ability to overcome shortages through effective management and use.

Access: This measures a country's ability to access water for drinking, industry and agricultural use. The 21 countries with very high scores are Austria, Barbados, Belgium, Canada, Croatia, Finland, France, Germany, Greece, Iceland, Japan, Netherlands, Norway, Portugal, Singapore, Slovakia, Slovenia, Sweden, Switzerland, United Kingdom and the United States. The lowest five are Eritrea, Chad, Ethiopia, Malawi and Rwanda.

Use: This measures how efficiently a country uses water for domestic, agricultural and industrial purposes. The lowest-ranking country is the United States, because of wasteful or inefficient water use practices. The U.S. also has high per-capita domestic water use, and high volumes of water are used per dollar of industrial production. Also in the bottom five are Djibouti, New Zealand, Cape Verde and Italy. The top five countries are Turkmenistan, Indonesia, Guyana, Sudan and Equatorial Guinea. They are relatively efficient in terms of the amount of water used, compared to revenue generated by that use.

Environment: This measures ecological sustainability, looking at water quality, environmental strategies and regulation, and numbers of endangered species. The top five countries in this category are Finland, Canada, United Kingdom, Norway and Austria. The U.S. is number six. The bottom five are Haiti, Morocco, Mauritius, Jordan and Belgium.

The situation in large Asian countries varies a great deal. While China, with its huge population topping a billion, scores quite well on capacity, and moderately on use, its scores on resource, access and the environment are all low. In India, a very low resource per capita score is counteracted by a relatively high score for use and capacity, but access and the environmental components are weak.

Canada will be a key player in moves to redress what Cosgrove calls the "virtual" movement of water through foodstuffs. "Canada has nine percent of all the world's fresh water, so that it can serve as a model of what a water-rich, wealthy country can do," he says.

Experts say that 20 percent of the world's population in 30 countries faced water shortages in the year 2000, a figure that will rise to 30 percent in 50 countries by 2025. They have warned that unless action is stepped up, the number of people living under threat of water scarcity will increase to 2.3 billion by 2025.

"Because the WPI includes indicators of health and of water quality, it can be used to address the link between lack of water access and ill health," says Dr. Sullivan. According to the World Health Organization (WHO), diarrheal diseases alone account for more than 3 million deaths per year, and give rise to 1 billion incidences of illness, many of which involve loss of capacity to work. Every year, more than 5 million people die from some kind of water-related disease, and more than 3 billion incidence of disease occur.

An estimated one half of people in developing countries are suffering from diseases caused either directly by infection through the consumption of contaminated water or food, or indirectly by disease-carrying organisms, such as mosquitoes, that breed in water.

"The WPI is work still in progress," says Dr. Sullivan. "It is not a definitive statement. The country rankings are not by any means the most important aspect of the WPI. It has been designed ultimately as a tool for monitoring progress, mainly at the community or district level." ❖

Questions

1. Who will attend the Third World Water Forum in Kyoto in March and what will they discuss?
2. The Water Poverty Index report grades 147 countries according to what five measures?
3. How much of the world's population faced water shortages in the year 2000 and what will that figure be by 2025?

Answers are at the back of the book.

12

Grappling with a dwindling water supply, the city of Atlanta must curtail its water consumption. In 2003, Georgia, Alabama, and Florida must come to an agreement on how to divide the water rights to the Apalachicola-Chattahoochee-Flint basin, which feeds all three states. With a water consumption rate that outpaces that of Las Vegas, Atlanta must find a way to satiate a booming population's thirst for water without leaving Alabama and Florida high and dry.

Atlanta's Growing Thirst Creates Water War

Douglas Jehl

New York Times, May 27, 2002

IT HAS ALL THE ELEMENTS OF A CLASSIC REGIONAL WATER WAR, pitting developers against environmentalists and state against state. Yet this battle is gripping not the parched Southwest, but the normally verdant Southeast, in a sign of future clashes around the country over an increasingly limited supply of fresh water.

Atlanta and its swelling suburbs, still ballooning with growth, rely for nearly all their water on the Chattahoochee River, a relative trickle of a waterway that is the smallest to supply so large an American city.

Until now, that dependence has not been a problem. Even in the last 10 years, as greater Atlanta's population soared nearly 40 percent, the withdrawals from the Chattahoochee have kept pace, with more than 400 million gallons now sucked from the river and a reservoir every day, helping to keep countless suburban lawns green.

But for the first time, Atlanta is being forced to admit that the current pattern cannot be sustained. That theme is at the heart of a dispute among Georgia, Alabama and Florida about dividing water rights for the next half-century, and it has left Atlanta to ponder what to do when its share of the Chattahoochee runs out.

With a June 17 deadline approaching for the governors of the three states to reach a deal, the dispute pits the growing thirst of Atlanta against the needs of downstream regions, including Apalachicola Bay, a pristine estuary on the Gulf of Mexico in Florida.

The decisions at hand may be the toughest on water that the Southeast has yet had to make, marking an end to an era in which abundant, cheap and barely regulated water has been seen as a kind of natural right in a region blessed by 50 inches of rain a year.

"In the past, water barely even entered into our calculations," said J. T. Williams, chairman of Killearn Inc., whose developments have added thousands of golf-course and clubhouse-community houses to the Atlanta area in recent years, with thousands more under way. But now, Mr. Williams said, "It's getting a little nervous for people in the development industry."

Georgia officials insist that they do not expect Atlanta to reach a real day of water reckoning until 2030, when they have projected that demands on the Chattahoochee will reach a maximum sustainable limit. But a recent draft report by the Army Corps of Engineers suggests that in some months, the

Chattahoochee may already be being tapped near capacity, a warning particularly alarming to Atlanta because its history and geology have left it with few good water alternatives.

"I think it's safe to say that water is now going to be the driving force in all of our decisions, and we're going to have to be a lot smarter about it than we have been in the past," said F. Wayne Hill, the top executive in Gwinnett County, one of the 16 counties in the sprawl of a greater Atlanta that now numbers more than 4.1 million people.

In Atlanta as in much of the East, the droughts of recent years have thrust water scarcity into greater public consciousness, through restrictions like the ones in place here, which allow outdoor watering only every other day. But public officials and experts here say that the multistate talks, with their focus on a 50-year future, have made clear the need to put in place in places like Atlanta more permanent but also more burdensome changes, in water pricing, regulation and conservation, along the lines familiar to the water-parched West.

"The idea that we're having water wars in a region that gets so much rain is astonishing, but it is definitely the shape of things to come," said Aaron T. Wolf, an expert on water conflicts and a professor of geoscience at Oregon State University in Corvallis. "The whole country is learning that we can't just keep on doing what we've always been doing when it comes to fresh water."

Unlike most American cities, Atlanta was founded, in 1837, not as a port but as a railway junction, far from any major river or lake. It rests on the hard rock of Piedmont granite, too impermeable to allow for the underground aquifers that in many parts of the country are the major source of fresh water.

In the last 10 years in particular, its growth has been dizzying, with the population of greater Atlanta climbing from 2.9 million in 1990 to 4.1 million in 2000, according to census figures. About 70 percent of the water supply is drawn from the Chattahoochee, including the waters blocked at Lake Lanier by the Buford Dam, built north of Atlanta in the 1950's by the Army Corps of Engineers, with the balance from a neighboring reservoir and river system.

In 1990, that meant that greater Atlanta pulled an average of about 320 million gallons of drinking water a day from the Chattahoochee, according to state and local officials. By 2000, with only slight declines in per capita water use, that figure had risen to 420 million gallons, and Georgia officials project that it will keep rising, to 705 million gallons a day by 2030, a point that even they describe as the maximum possible, given other demands on the river, including hydropower and the need to sustain ecosystems downstream,

particularly in the Apalachicola River in Florida, which carries the waters of the Chattahoochee 109 miles to the Gulf of Mexico.

But the question of exactly how to divide the rights to the river, in what is known as the Apalachicola-Chattahoochee-Flint basin, has stymied the governors of Georgia, Florida and Alabama since they were entrusted with the task by Congress in 1998.

As the talks now stand, an insistence by Florida on measures designed to guarantee sufficient year-round flows into the oyster- and shrimp-rich Apalachicola remains confronted by vehement opposition from Georgia to what it regards as thinly veiled proposals for the micromanagement of water it has always regarded as its own.

"We need to be able to use our own water resources that were built and planned for this area," said Pat Stevens, chief of environmental planning for the Atlanta Regional Commission. "I just don't see any alternative."

"Left unconstrained they'll suck it dry," said David McLain, executive director of Apalachicola Bay and Riverkeeper, in Eastpoint, Fla., in a reference to the politicians and developers of greater Atlanta. "What we'll be left with is a muddy ditch."

The June 17 deadline is only the latest in a series that have so far come and gone with little more than an agreement by the three states to extend the talks. If the negotiations break down, the dispute will almost certainly end up at the Supreme Court, which could wrest any solutions out of the states' control.

But while the governors of the three states say they would prefer to resolve the matter themselves, none is likely to sign on to a deal unless he can portray it as the best for his residents. All three governors are up for re-election this year, intensifying a shared sense of how much is at stake.

"Take the slogan 'Gov. Fill-in-the-Blank is giving our water away' and you have a kiss of political death," said George William Sherk, a water lawyer who teaches at George Washington University in Washington.

The fight over water allocation has links to other disputes in the river basin, including a longstanding debate about the dredging of the Apalachicola, a multimillion-dollar a year task by the Army Corps of Engineers to make possible navigation by no more than a handful of barges. Sand and silt from the dredging have clogged the swamps that line the river and that are home to large numbers of endangered plants, reptiles and amphibians.

Despite strong opposition from some lawmakers, including Representative Bob Barr, Republican of Georgia, and Senator Bob Graham, Democrat of Florida, Congress appropriated new money for the dredging last year.

The dredging maintains a nine-foot deep channel in the Apalachicola River from the Gulf of Mexico to the Jim Woodruff Dam at the Georgia state line, allowing barge traffic to reach the Georgia port city of Bainbridge. Just 47 barges moved up the river from the Gulf of Mexico in 1989, the most recent year for which figures are available.

Of the three states, only Georgia relies on the rivers in the basin for large supplies of fresh water; Florida, short of water elsewhere, has plentiful supplies for its sparse population in the area where the Apalachicola flows into the sea. In general, Alabama has aligned with Georgia, leaving Florida to make the case to give the needs of downstream ecosystems as much weight as those of upstream populations.

In the talks, Florida has so far demanded that Georgia sign on both to agreements that would guarantee minimum flows of water at the point where the Chattahoochee crosses the Florida line, and to limits that would affect how much river water Georgia could withdraw. Georgia has said it could agree to one or the other, but not both, saying it will need flexibility in order to provide the water that Atlanta needs.

But as a sign of a new era, state and local officials are now talking openly about the necessity for bigger changes in water use if Atlanta's rapid growth is to be sustained. These would include seeking out new water supplies, but would focus on price increases and other regulations designed to sharply reduce municipal water consumption that averages about 160 gallons per person per day, far higher than in places like Phoenix and Los Angeles.

"When it comes to water, most of the folks in Atlanta have been fairly spoiled over the years," said Ted Larrabee, executive coordinator for the Metropolitan North Georgia Water Planning District, established last year to oversee all of greater Atlanta. "There has been a lack of programs that do anything more than give lip service to conservation, and that is what has to change." ❖

Questions

1. When do Georgia officials expect Atlanta to reach a real day of water reckoning?
2. What was Atlanta founded as and why is this significant?
3. How much of greater Atlanta's water supply is drawn from the Chattahoochee?

Answers are at the back of the book.

13

Biodiversity, or the number of species inhabiting the earth, is sharply declining. Loss of habitat is considered to be the biggest contributor to the earth's decline in biodiversity. With rising pollution, habitats are becoming scarcer. The degradation of lands and waters forces wildlife into areas that are unsuitable to live in, thus pushing their populations ever closer to extinction. According to a recent study, half of the most biologically diverse areas in North America are considered to be severely degraded. A major concern is that there is a lack of environmental policy coordination for the cross-boundary areas of the United States, Canada, and Mexico.

North America Losing Biodiversity, Say Experts

Danielle Knight

Global Information Network, January 7, 2002

LOSS OF HABITAT FOR FLORA AND FAUNA in Canada, Mexico and the United States has caused a "widespread crisis" of shrinking of biodiversity throughout the region, an inter-governmental environmental body warned today.

The remaining natural environments in the region are under enormous stress and have been fragmented, polluted, or damaged in other ways despite efforts to set aside protected areas, the North American Commission for Environmental Cooperation (CEC) said in a report to the three countries' governments.

"Over the past few decades, the loss and alternation of habitat has become the main threat to biodiversity," said Janine Ferretti, executive director of the Montreal-based CEC, which was set up under the 1993 North American Free Trade Agreement (NAFTA) at the request of environmental groups.

The report, "The North American Mosaic: A State of the Environment Report," was based on scientific papers prepared for the CEC by scholars and government experts in the three nations.

Half of North America's most biologically diverse areas were severely degraded, said the report. At least 235 species of mammals, birds, reptiles and amphibians—including the Monarch Butterfly and Northern Codfish—were threatened.

Freshwater species were especially at risk because physical barriers, including dams, prevented them from escaping to new ecosystems when their own habitat became destroyed or degraded.

"The United States contains the world's greatest diversity of freshwater mussel species but more than 65 percent of these are threatened or extinct," said the report.

Fragmentation and loss of forest mean that many migratory birds, which depend on healthy, contiguous forests, "are losing nesting, feeding, and resting areas," it added.

In 1999, environmental groups in the three countries filed a formal complaint with the CEC, arguing that the United States was flouting international and domestic protections for migratory bird species.

Advocacy organizations accused U.S. officials of having an unwritten policy of not investigating cases in which migratory birds were killed by logging operations. Tens of thousands of migratory birds are killed each year as a result of road building, cutting, bulldozing, and burning, they said.

Similarly, the monarch butterfly, an insect whose migratory path spans the three nations, faced habitat threats, according to the CEC report. These included coastal development in California, deforestation of oyamel fir forests in Mexico, and the use of pesticides on and around milkweed plants, the butterfly's primary food.

The report said all three countries had taken action to protect biodiversity, such as increasing nature reserves. Mexico, for example, has created 10 new "biosphere" reserves in the past ten years, while Canada's total area of protected land has tripled since 1970.

The total protected area in North America has increased by from less than 100 million hectares in 1980, to 300 million hectares, or about 15 percent of the continent's land surface.

Looming threats to protected areas, however, overshadowed these positive achievements, according to the CEC. "Natural areas in all three countries are in danger of being overwhelmed by multiple factors," it said.

Reserves and other protected areas did not have sufficient management funds and were harmed by an increasing amount of visitors. Land surrounding parks and preserves were often developed and threatened the survival of the protected area, it said.

The report also warned that Canada and the United States were putting additional strains on the environment because they consumed more fresh water per capita than any other country in the world—using about twice as much per person as Mexico.

The heaviest demand for freshwater came from agriculture and thermoelectric power generation, which together accounted for about 80 percent of water withdrawals in the two countries.

A large share of irrigation water was pumped from underground aquifers created by the accumulation of small amounts of rain over many centuries. One of the world's largest aquifers, the Ogallala aquifer that lies under the Great Plains in the United States, for example, was being depleted at a faster rate than it was recharged, it said.

For the most part, soil loss through erosion by wind and water was decreasing due to better conservation practices and programs, according to the report. Between 1982 and 1997, total erosion on all cropland in the United States decreased by 41 percent, from 3.08 billion tons in 1982 to 1.81 billion tons in 1997.

As a result of implementing reduced tillage agricultural practices in Canada, the risks of wind erosion in the county's prairies dropped by 30 percent between 1981 and 1996, said the report.

Although soil erosion declined in many parts of North America, the report warned that more soil was being lost in agricultural areas than was being regenerated.

"Part of the problem is a lack of humus because of a heavy reliance on chemical fertilizers rather than on traditional fertilizers and soil amendments, such as manure and compost, that help maintain soil structure," it said.

The report said that low-income communities, especially indigenous groups in the region, were hardest by environmental problems. Some of the wildlife and plants that are an important part of the diet of native communities in the Great Lakes basin and in the Arctic were unsafe to eat because they were contaminated with pollution.

"Across the continent, forest clear-cutting, unsustainable resource extraction, industrial pollution, and over-fishing often take the greatest toll on low-income communities," said the report.

The report concluded that the three NAFTA nations did not adequately coordinate efforts to address cross-boundary environmental issues, including cross-border habitats, non-native species, and shared migratory species and water resources. None of the three nations have adequate methods of compiling pollution statistics, it said.

"There is an urgent need to develop mutually compatible economic, social and environmental goals and policies across the three- country region," it said. ❖

Questions

1. Why are tens of thousands of migratory birds killed each year in the United States?
2. How much has total protected area in North America increased?

3. Why has soil loss through erosion decreased?

Answers are at the back of the book.

14

Much of British Columbia's coastal forests have fallen victim to deforestation. Areas as large as several hundred acres may be completely leveled because of the practice of clear-cutting, which accounts for 95% of the logging in British Columbia. While some measures have been taken to protect parts of the Great Bear rainforest, old-growth forests are being mowed down by British Columbia's timber planners. Operating under the theory of "sustainable yield," old-growth forests are being cut in massive amounts in hopes of speeding their regeneration. This practice is being challenged by the Northwest Ecosystem Alliance, other conservation groups, and the U.S. timber industry on the grounds that the Canadian timber industry is being effectively subsidized and has an unfair competitive advantage in the world market.

Buzz Cut

Paul Rauber

Sierra, April 2003

THE POWERFUL WESTERLY THAT BLEW DOWN from the icy summits of British Columbia's Coast Mountains in the middle of the night threatened to send our tent and everything in it into Shack Lake. "At least it's dry," I said to my wife. That's when the rain started. By morning we were covered by half a foot of snow, and resigned ourselves to tentbound hibernation.

If you're going to get snowed in on an autumn camping trip, you could find worse company than Canadians. As we sat shivering in our tent, our British Columbian companions sprang into action, cleaning out the bear-ravaged miner's cabin that gave Shack Lake its name. Soon we were drying our sodden woolens over an ancient cast-iron stove and sharing stories of wretched outdoor ordeals. Veteran wilderness campaigner Gil Arnold told of a mid-March excursion to the ominously named Tombstone Lake in Alberta's Rockies, where for days on end the thermometer hit 30 below with 30-mile-an-hour winds. The trip included both hardy plaid-clad Canadians and pile-and-Gore-Tex-sheathed Americans, hunkered down in their respective tents. At one point, one of the Americans came scratching at the Canadians' tentflap. "What are you eating in there?" he inquired. "Bacon, bannock, and

rum," they replied. "What are you eating?" "Freeze-dried death," replied the miserable Yank.

Luckily for us, the Canadian conservationists on our trip to the Chilcotin region of southcentral British Columbia were also of the rugged frontier variety. Most seem to have homesteaded at some point in remote Yukon cabins, setting traplines and mushing sled dogs. When Arnold stumbled on a hike and sliced his thumb open on a sharp rock, he patched it up with duct tape. (Probably made it easier to stir his coffee.) On our way to Shack Lake we had stayed at the self-built, off-the-grid home of Dave Neads, a campaigner with the tiny environmental powerhouse BC Spaces for Nature, and his wife, Rosemary; our departure was delayed when Neads had to repair his water line, which had been chewed through by a bear.

When practiced by the Canadian timber industry, however, that same frontier spirit proves environmentally disastrous, just as it has with U.S. timber companies in these denuded United States. British Columbia's forests are so vast that they've been treated as endless—with the inevitable result that they are ending. For example, half of the trees in the Chilcotin region (an area that covers 15 percent of British Columbia) were cut between 1988 and 1996 alone.

Clearcutting accounts for 95 percent of the logging in British Columbia, with "cutblocks" as large as several hundred acres.

We had seen scores of these a few days previous, flying out from Nimpo Lake in a Beaver, the workhorse plane of the north, piloted by a veteran named Floyd. ("He's been flying for forty years," Neads reassured Joe Scott of the Northwest Ecosystem Alliance, a burly outdoorsman but an unhappy flyer. "That's a good sign," said Scott hopefully. "But then," continued Neads mischievously, "there's an increased chance of heart attack . . .")

From the air, the clearcuts are irregular gray jigsaw pieces in a sea of lodgepole pine, with only the skimpiest fringe separating them from lakes, wetlands, and streams. British Columbia requires unlogged streamside buffers one-fifth to one-half the size of those in U.S. national forests—and none at all for streams less than five feet across.

In British Columbia, mighty logging firms like Weyerhaeuser and International Forest Products (Interfor) are free from a host of other environmental restrictions that rein in their colleagues to the south. Canada, for instance, has no Endangered Species Act. Cut levels on crown lands (which are owned by the government and hold 95 percent of Canada's timber) may be restricted to protect biodiversity and some wildlife habitat—but not by more than 4 percent and 1 percent, respectively. Even in "high biodiversity" areas, British Columbia's logging plans foresee leaving only 10 percent old growth over the long term. And timber companies can't cut less even if they wanted: To protect timber and mill jobs, the province requires a minimum harvest, regardless of market conditions.

Political conditions, however, are starting to change. This April, thanks to a multiyear campaign by the Sierra Club of British Columbia and other environmental groups on both sides of the border, British Columbia agreed to prohibit logging on 1.5 million acres of the coastal Great Bear rainforest, and to defer logging on an additional 2 million acres until a more environmentally sensitive forest-management plan is completed. The Great Bear, one of North America's largest remnants of ancient temperate rainforest, boasts mist-shrouded stands of 800-year-old cedar, roiling salmon streams, prowling grizzly, and the ghostly white kermode or spirit bear (see "Canada's Forgotten Coast," March/April 1999).

The Great Bear decision was a tremendous win for conservationists, but beyond such fortunate protected areas, British Columbia's forest policy still remains one of liquidation and conversion to tree farms. In the Chilcotin, a Switzerland-size plateau on the dry, inland side of the Coast Mountains, forests of Engelmann spruce, Douglas fir, and lodgepole pine are systematically mowed down. "The interior forest is forgotten forest," says Neads. He got his own stark reminder several years ago when he and Rosemary noticed eerie bright lights in the hills south of their remote cabin: The clearcutting didn't even stop at night. Seventy-five percent of British Columbia's cut comes from the east side of the Coast Mountains, much of it from the Chilcotin, and 90 percent of that goes to the United States, half as wood chips for paper, half as two-by-fours.

Most of the 125- to 250-year-old pines in the Chilcotin aren't big enough for anything else. While the Coast Mountains can receive up to 60 feet of snow a year, the Chilcotin gets only 10 to 15 inches of annual precipitation and 30 frost-free days a year. With such a short growing season, 200-year-old trees look like saplings. (In some parts of the Chilcotin, the volume of wood per acre is the same as that of a single tree on the coast.) Visiting the enormous log decks at the mill at Nimpo Lake, a tourist resort ringed with clearcuts, I didn't see a single log more than a foot in diameter. And in the mournful stubble of a four-year-old clearcut nearby, there was no sign of replanting or regeneration, just the violent excretions of the feller bunchers, massive tractors that grab the trees, snip their bases with huge shears, and load them onto waiting skidders.

In the wishful thinking of British Columbia timber planners, this block will be cut again in 80 years. In the theory of "sustainable yield," clearcut forests will regenerate themselves by then, providing another "crop" of trees. To speed this process, the province is hewing its virgin forests as fast as possible. "There is no second-growth cutting in the Chilcotin," says Neads. "All of the forest being cut is virgin old-growth."

Our chill mountain camp on the Chilcotin side of the Coast Mountains lay near the treeline, where clearcut gave way to krummholz, the low, twisty trees of the alpine tundra. Skirting permanent snowfields and flushing white-tailed ptarmigan, we labored up the mountain that rose 1,700 feet above us, stopping occasionally to admire the massive racks left by woodland caribou or to graze on the profusion of ground blueberries at our feet. Naturally the Canadians were in the vanguard, and we heard serial exclamations as each party reached the summit and glimpsed the amazing panorama before them. The central Coast Mountains filled our field of vision from the distant Bella Coola valley in the north to the heights of Waddington, Monarch, and Pagoda, with the deep trench of the Klinaklini River cutting through in the south. Far off in the river's upper reaches in Hidden Valley we could see small puffs of dust—the logging trucks hauling the forest away to the mill at Williams Lake, 14 hours distant. (The coroner there is trying to put a stop to one-day round trips because of the high number of highway fatalities.)

At this time last year, BC Spaces for Nature was waging a desperate campaign to save the magnificent but little-known

Klinaklini, one of the few rivers to cut through the rock wall of the coastal ranges to drain the interior. Its nearly 60 miles of riparian wilderness were being nibbled at either end by roads and chainsaws, even though the road in its tortuous lower reaches cost $1 million per kilometer to build—an expense that the province allowed Interfor to charge against the already minimal "stumpage" fee it paid for the trees. Thus Interfor pays only 25 cents (Canadian) for a cubic meter of wood instead of the more usual rate of $30. (By U.S. measures, this works out to 26 board feet for a penny.)

A few days later we piled into a float plane to survey the Klinaklini, named "the river that winds back on itself" in the language of the indigenous Chilcotin people for its snaky path. (The less poetic coastal Kwakwaka'wakw called it "eulachon grease" after the fatty smeltlike fish that crowded its mouth.) Steep, heavily wooded slopes and towering waterfalls rushed down to the river bottom, where the sparkling flood mirrored the mountains and sky above. Beaver engine droning, we fell into the reverie that magnificent country inspires: "Next summer, let's visit that valley," "I bet there's trout there," "Wow! Could we kayak that?" (The answer to the latter is no: At one point the Klinaklini enters an impassable canyon that requires a helicopter portage.) We passed Dorothy Creek, the inland terminus of an approved logging allotment, and were suddenly jarred to attention by the sight of brilliant red and yellow feller bunchers below, tearing the guts out of the wild forest.

Earlier this year, the destruction of the lower Klinaklini ended, thanks to the Great Bear agreement. The area up to Hidden Valley, where we had seen the dust of logging trucks, has been protected, although a late challenge is coming from mining companies, which may try to scuttle the deal with the province's new ruling Liberal Party (which, despite the name, is more akin to the Republican Party in the United States).

The logging continues in Hidden Valley, but that too may come to an end as a result of an unusual challenge under the same international trade rules more commonly used to weaken environmental protections. The Northwest Ecosystem Alliance and other conservation groups, in an odd coalition with the U.S. timber industry, are arguing that British Columbia's lax environmental rules and below-market stumpage fees are, in effect, subsidizing its timber exports, giving them an unfair competitive advantage. "There's too much wood being cut," says Scott. "The wood is subsidized, so they cut large amounts, which tends to depress prices because the supply is so great. The lack of environmental protection in Canada is actually hurting U.S. industry." The U.S. Department of Labor has certified that Canadian imports have been a factor in the closing of more than 50 U.S. mills since 1996. With the support of the U.S. International Trade Commission, the U.S. coalition is asking for duties as high as 78 percent on lumber imported from Canada.

At press time, the U.S. Commerce Department was set to decide whether to impose the tariffs. If it does, Canada will surely appeal the case to a special NAFTA court composed of two U.S. and three Canadian judges, which could overrule the Commerce Department and reverse the duty. Based on the court's rulings in similar cases in the past, Canada is confident that the NAFTA panel will bless the subsidized liquidation of its wild forests.

But those were battles for another day. If we were to engage in them, we'd have to leave Shack Lake, but the prevailing down-mountain winds threatened to delay our departure indefinitely. Maybe tomorrow, Floyd said. Tomorrow came, and though the wind was still kicking up, Floyd told us to get in the plane anyway, but to leave our gear behind. We taxied up to the head of the lake, turned around and roared back, but had to abort when we were three-quarters down the lake and our skids hadn't left the water. I began to wonder how many days it would take to walk out through the new snow.

Then Floyd turned back around toward the head of the lake and opened up the throttle. At first I thought he meant to take off in that direction, but it was obvious that he couldn't possibly gain enough altitude to clear the mountain. (Another Beaver had crashed a few weeks prior in a failed downwind turn in a similar situation.) Then I thought he was just in a hurry to get to the head of the lake for another run. Only when we were turning around again at full throttle up on one pontoon did I realize that he was buying extra speed at the start to get us off the water, barely clearing the treetops at the bottom of the lake.

It was more exhilarating than frightening. For forests and their defenders alike, there is nothing quite like survival. ❖

Reprinted with permission of *Sierra*, the magazine of The Sierra Club.

Questions

1. Thanks to a multiyear campaign by the Sierra Club of British Columbia and other environmental groups, how are political conditions starting to change in British Columbia?

2. Where does 75% of British Columbia's cut come from and how much of it goes to the United States?

3. How have Canadian imports affected U.S. mills?

Answers are at the back of the book.

While the earth's food supply and demand are currently growing at the same rate, it has been projected that eventually demand will outpace supply. In spite of this ominous prediction, some evidence has shown that a global food crisis can be avoided. Resource depletion and environmental degradation, however, are factors that could alter the current supply and demand projections. As some of the world's nations struggle through their developing phases, they experience greater resource depletion and environmental degradation than most developed countries. The predicament is whether or not developing nations can pass through their more environmentally destructive developing stages without permanently injuring the environment.

Feeding the World

Luther Tweeten, Carl Zulauf

The Futurist, September/October 2002

IN OCTOBER 1999, THE EARTH'S POPULATION surpassed 6 billion people, a milestone greeted by doomsayers with expressions of gloom. Supporting this gloom is the declining growth rate in crop yields. Though supply and demand are now growing at almost the same rate, there are still fears that food demand will eventually outpace food supply. Growing evidence shows, however, that when negative population growth predicted by the end of the twenty-first century-as well as environmental and other factors are taken into account, we can avoid a threatened global food crisis.

Society's Four Transitional Stages

Economists often refer to four transitions or stages of societies based on demographics, economics, agricultural productivity, and technological advancements. The first stage is a traditional society, characterized by low population density and low economic and population growth. In this society, which may be a country or a region, high birthrates match high death rates, while primitive technology contributes to low income and low living standards.

The second stage is a developing stage, occurring when a society's technological advancements result in sustainable agricultural production and plant and animal domestication. The result is a more plentiful food supply, which helps increase population growth mainly by slowing death rates. When population and food production growth are combined with industrialization and urbanization at this stage, the result is environmental exploitation and degradation present in many developing societies.

Increases in agricultural productivity and production bring economic surpluses that allow growth in capital and per capita income; the third stage, the developed society, is born. Birthrates fall faster than death rates as the roles of women change and developments in birth-control methods allow adults to choose the number of children they want.

Finally, stage four, the mature society, sees notable technological change evolve beyond agriculture (particularly in medicine and public health), making death rates decline further.

While many technological breakthroughs in developing societies come from innovative laypersons, breakthroughs in developed societies tend to require scarce, highly trained, experienced, and costly technicians and scientists. By the time the mature society develops, the most readily accessible raw materials have been exploited. Obsolescence of current technology requires investment in maintenance rather than in new technologies. Increase in productivity of service activities, which grow in importance, becomes more difficult than increases in agricultural and manufacturing productivity. Some developed regions choose to sacrifice some economic growth for equity. Thus, while productivity and income continue to rise, the rate of these increases slows.

Rapid agricultural productivity gains continue in developed societies as investments in education and science made in the development stage produce long-term payoffs and as urbanization and industrialization lead to an exodus of agricultural labor. At the same time, slowing rates of income growth and population growth also slow down the growth in demand for food. Food self-sufficiency increases in some countries after falling in the development stage. However, agricultural trade typically grows as more affluent consumers demand a variety of foods from around the world.

Many developed countries have recently entered or soon will enter the fourth stage-the mature society-the future society of the world's inhabitants. The long-held view is that global population growth will more or less stabilize; recent evidence, however, presents a strong case for negative global population growth as the seminal attribute of the mature society.

Evidence for Negative Population Growth

According to the United Nations, the total fertility rate-the number of children a woman may be expected to bear during her lifetime-has fallen in every region of the world since 1950. From an average of nearly six children per woman in the 1950s, by the early to mid-1990s total fertility rate fell to three children in Latin America, 3.4 in India, and 3.5 in other parts of Asia. The only major exception to this sustained downtrend is in North America, where the recent increase in total fertility rate appears to be a transitory phenomenon associated with immigration and a large number of baby-boom women deciding to have children relatively late in their lives.

As significant as declining total fertility rates worldwide is the fact that, from 1990 to 1995, these rates in Europe, China, and North America were below the 2.1 average children per woman needed to sustain population worldwide over the long run.

The United Nations' medium population projection of 2.1 children after 2040 is widely used as a demographic forecast, but it unrealistically assumes that this rate will be the same in both developed and developing countries. Many researchers, such as Wolfgang Lutz of the International Institute for Applied Systems Analysis, do not support the UN assertion that fertility would increase to replacement level in developed countries. Lutz and others cite evidence pointing toward low fertility, noting contraception, declining marriage rates, high divorce rates, increasing independence and career orientation of women, materialism, and consumerism.

"These factors, together with increasing demands and personal expectations for attention, time, and also money to be given to children, are likely to result in fewer couples having more than one or two children and an increasing number of childless women," Lutz and his colleagues write in The Future Population of the World (Earthscan, 1996).

The United Nations has a second scenario-the low/medium scenario that presumes fertility averaging 1.9 children per woman for all regions by 2025. This scenario may be as unrealistic as the medium population scenario. The low/medium scenario may underestimate future total fertility rates in developing countries, just as the UN medium scenario may overestimate future total fertility rate in developed countries.

The low/medium scenario projects a peak world population of 7.9 billion people in 2050, declining to 6.4 billion by 2150. The medium scenario projects a peak world population of just less than 11 billion by approximately 2200. Most other projections, however, predict peak global population in less than a century, followed by negative population growth.

These 1998 UN population figures were revised in 2000, and the new estimates, though tentative, indicate population trends even lower than the 1998 predictions. The low/medium scenario is comparable to what the United Nations now calls the low variant for worldwide populations. Given the low variant, the United Nations predicts that world population in 2050 will be 7.8 billion—slightly less than the 7.9 billion projected under the low/medium scenario. We predict that the United Nations will continue to revise population trends downward and that negative population growth will occur even sooner than by 2150. However, because the 2000 population data remain tentative and do not extend to 2150, we prefer to continue to use the 1998 data for the remainder of this article.

Future Food Supply/Demand Balance

World food production increased 2.3% annually from 1961 to 1999 (see graph). Moreover, growth in world food production outpaced growth in population. As a result, average per capita food production increased by 0.5% annually while real food prices declined by 1.8% annually.

These trends mask changes over time in the contribution of yield and area. From 1961 to 1970, global food yield rose while cropland area stayed constant. From 1970 to 1990, both yield and cropland area rose. Since 1990, cropland area has stayed nearly static while yield continues to increase. Millions of cropland hectares added in Brazil and other countries during the 1990s were offset by millions of cropland hectares abandoned in the former Soviet Union as well as by the conversion of cropland to urban and other noncrop uses around the globe.

The finding that world food output during the 1990s depended on increased yields is important because food demand has been increasing as fast as yields since 1961. Growth in food demand takes into account not only growth in population but also growth in per capita consumption as incomes rise.

A world of no increase in acreage and similar increases in food demand and yield raises the potential of ominous outcomes. The "Alternative Scenarios for Global Food Supply and Demand" graph shows alternative food demand projections from 2000 to 2150 as well as projected food supply based on a continuation of 1961–1999 yield trends with no increase in crop area. If population and income continue to grow at the same rate as they grew from 1995 to 2000, the demand for food will sharply outgrow future food supply. Real commodity prices would need to rise to draw additional land and other resources into food production and to restrain consumption. If the population grows at 2.1 children per woman (the medium UN projection), food demand growth will outstrip food supply growth, but only until approximately 2075. This gap could probably be covered by small increases in real prices. If population growth follows the UN low/medium population scenario (1.9 children per woman), food supply will grow faster than food demand. This outcome would allow real farm commodity prices to continue to fall. It follows that the trend toward negative population growth may only narrowly avert rising real farm commodity prices in the first half of the twenty-first century.

Protecting the Environment

Natural-resource depletion and environmental degradation could confound the foregoing food supply and demand projections. Few activities devastate the environment more than dense populations of poor, hungry peasants clearing and cultivating rain forests and other biodiverse lands for food.

Environmental degradation is minimal in the traditional society but increases in developing societies as population and economic growth accelerate. Sustained population growth leads to land clearing and soil erosion that accompany food production expansion. Although economic growth occurs, developing countries do not possess the financial resources to pay for costly pollution controls or conservation technologies. Low-income consumers prefer present consumption without taking into account future considerations, further discouraging investment in environmental protection. Streams become polluted and air quality deteriorates, two of the more

obvious manifestations of wide-ranging environmental degradation as a society moves from developing to developed.

Several studies have shown that the interaction between population and income will ultimately save the environment while leaving sufficient land for food production. These conclusions assume that rising affluence is attended by effective policies to establish appropriate environmental care incentives and institutions. First, higher income in part brings about lower population growth, which reduces pressure on the environment. Second, as incomes rise, consumers spend a larger share on services, production and disposal of which are generally less detrimental to the environment than other forms of consumption. Higher-income consumers also demand greater efforts to protect the environment because, once their basic needs are met, they become more concerned about environmental quality and have the income to do something about it. Education and research made possible by economic progress also promote awareness of the environment, which in turn can generate effective policy responses. The net result is increasingly less environmental damage, followed by environmental preservation.

Higher income also means that current earnings are no longer needed solely to provide necessities and can be used to finance investment in science and technology. Higher per capita income and slower population growth contribute to lowering harmful emissions into the environment and reducing the use of natural resources. And rapid gains in agricultural productivity made possible by research enable farmers to cultivate fewer acres, freeing arable land for grass, trees, recreation, and biodiversity. Ultimately, though we will use less acreage for crops, we will have more green, environmentally friendly space. And because food yield per acre has increased, there will still be enough food production.

The Challenges Ahead

Interacting demographic, economic, and environmental transitions produce mostly positive outcomes. Of course, making predictions with absolute certainty is a dangerous undertaking; unforeseen factors can intervene-and probably will. What follows are other considerations that further affect or accompany food supply and demand issues.

The United Nations' low/medium population projection suggests that food supply growth will continue to outpace food demand growth, resulting in continuing decline in real prices for farmers. Falling fertility rates will avert a global food crisis. However, the optimistic projection of the food supply and demand balance should not be a basis for complacency; commitment to public and private investments in food science and technology will be essential to avoid future real price increases in food.

The declining real price of farm commodities will spur development of new nonfood uses, including energy, plastics, and pharmaceuticals. It is conceivable that these uses will exceed food uses in the not-too distant future. Nonfood uses will make for a more price-elastic demand, thereby diminishing perennial problems of price instability in agriculture; that is, farm commodity prices will become less volatile.

Trade must also be taken into account. From little or no international trade in the traditional society, agricultural and general trade increases in developing societies and remains high once a society is developed and higher incomes and population growth encourage food imports. While demand growth will fall with negative population growth, thus slowing trade in mature cultures. However, food businesses will still drive trade among nations to achieve economies of size and meet specialized needs. Furthermore, for food products, consumers in mature societies make their choices based more on lifestyle considerations than on prices. Thus, agricultural trade in a mature society remains high to supply the diversity of food products demanded by high-income consumers.

Africa's population will continue to increase, even under the United Nations' low/medium scenario, approaching 2 billion people in 150 years. Local food production is not expected to keep pace with growth in food demand. By 2020, cereal imports by sub-Saharan Africa could be triple the 1990 level. Building buying power within Africa to purchase needed food imports will be challenging indeed.

The low-input, sustainable agriculture of the traditional society contrasts sharply with environmental deterioration at an initially low but increasing rate in developing societies, and at a high though decreasing rate in developed societies. In a mature society, however, the accumulation of wealth, science, technology, and knowledge coupled with declining population permits a turnaround in natural-resource depletion and environmental degradation. Thus, mature societies provide optimism for the environment. The environmental dilemma is how and whether developing countries, including China and India, can pass through the developing and developed stages of high environmental degradation and natural-resource depletion to the higher per capita income of their maturity-when environmental preservation predominates -without irreversibly damaging the global environment in the process. Assuming appropriate policy is followed, falling world population plus economic development will allow this transition to occur.

International migration involving both high- and low-skilled workers will become pervasive. Farms in North America and Europe will continue to seek immigrant labor to perform labor-intensive jobs. Migration will lead to greater cultural diversity within nations and could cause a narrowing of cultural gaps. The potential for cultural and ethnic strife will be real, however, and could negatively affect food production and trade.

The exodus of labor from agriculture will continue. Negative population growth will lower labor supply and raise the relative cost of labor, as well as of management and entrepreneurship. Technological change will replace labor and human intelligence, at least at lower cognitive skill levels, with computer intelligence and robotics.

Ethical questions accompany many of these issues, especially in countries with negative population growth: How far should machines be allowed to replace workers? How far should genetic engineering be allowed to alter plants and animals? While these questions will energize policy debates across all segments of society, agriculture's place as a producer of food and the strong emotions that it engenders as a way of life suggest that this debate will be especially energetic when it involves appropriate policies for agriculture. ❖

World Population Distribution

Under both the United Nations' medium and low/medium population scenarios, large demographic realignments will occur among the world's regions. For example, under the low/medium population scenario, the United Nations predicts that the population in Europe, North America, and countries in the Pacific (Australia, New Zealand, and Micronesian and Melanesian nations) will significantly decline, from 29% of the world's inhabitants in 1950 (732 million people) to nearly 10% (615 million people) by 2150. Europe's population is expected to drop to half its current level during this time period. At the same time, Africa's population will grow from nearly 9% (224 million) in 1950 to nearly 30% by 2150 (1.8 billion), while India will become the most populous country by 2050, supporting 1.3 billion people.

High incidence of AIDS in parts of Africa complicates making population projections for this continent. J. Bongaarts in his essay "Global Trends in AIDS Mortality" projects that AIDS will reduce the population growth rate in sub-Saharan Africa to 2.6% per year by 2005, compared with 2.8% per year without AIDS in the same region. The impact would be even greater except that many adults bear children before succumbing to AIDS. Nonetheless, Africa's population growth rate could fall further than anticipated unless the AIDS epidemic is brought under control.

—Luther Tweeten and Carl Zulauf

Genetically Modified Crops

- Estimated global area of genetically modified (GM) crops is 130 million acres (52.6 million hectares) in 2001, up by 19% from 2000.
- Principal GM crops:
  Soybeans: 83 million acres (63% of global GM crop area).
  Corn: 24 million acres (19%).
  Cotton: 17 million acres (12%).
  Canola: 6 million acres (50%).
- Approximately 68% of GM crop area is in the United States (89 million acres).
- More than 25% of GM crop area is in six developing countries: Argentina (29 million acres), China (3 million acres), South Africa (500,000 acres), and Mexico, Indonesia, and Uruguay (less than 250,000 acres each).
- Other countries growing GM crops include Australia, Bulgaria, France, Germany, Romania, and Spain.

Originally published in the September/October 2002 issue of *The Futurist*. Used with permission from the World Future Society, 7910 Woodmont Avenue, Suite 450, Bethesda, Maryland 20814. Telephone: 301/656-8274; Fax: 301/951-0394; http://www.wfs.org

Questions

1. What was the earth's population in October 1999?
2. From 1961 to 1999, world food production annually increased by how much?
3. Even under the United Nations' low/medium scenario, what will be Africa's population in 150 years?

Answers are at the back of the book.

Dealing with Environmental Degradation

Small rural towns, such as Buffalo, Illinois, are commonly overlooked by societal standards. Many of these towns have become toxic wastelands and dumping grounds for industries. Unfortunately, citizens of these small rural towns frequently experience higher rates of cancer than other towns. Oftentimes wildlife die-offs occur first as a precursor of human illnesses and deaths. The blue-collar workers of these towns are often asked to handle toxic and carcinogenic chemicals under the threat of losing their jobs, while "mums the word" on many such stories.

Ill Winds: The Chemical Plant Next Door

Becky Bradway

E: The Environmental Magazine, September/October 2002

FROM THE TIME I WAS EIGHT until I was 10, I lived in a pea-sized town down the road from the chemical plant and the munitions factory. Buffalo, Illinois had a bank, a grocery store, a post office and a park. And enough houses to hold three hundred people. This was where we moved, a few miles from my grandparents, after we left Phoenix. My mother never wanted to come back to Illinois. Once she was there, she only spoke to the relatives, and when they were with us they seemed to mask the hole in her life.

I loved that town nearly as much as my mother hated it. She hated it so much that she never went outside except to get in her car and drive away. She sent me to the post office and the grocery store, had my dad do the yard work. She sat in the dark living room watching soap operas and folding clothes. The kitchen was for packing my father's suppers. He worked swing shift—3 to 11 p.m.—or graveyard, the night shift. When he was gone, we would eat Kraft Macaroni and Cheese and carrot sticks. When he was home, we would have pork chops or the end of a roast. We tried not to spill the milk or sing at the table.

Anyway, it was the town that mattered. A town, or what it should be, is a community of hope. It's bustle and defiance out here in nowhere. Buffalo was my sanctuary, my circle. Friends, school, church, street, sidewalk, lawn, bells. They were not artificial constructions. They were not just institutional and conformity-approving props. We need these places. I needed them.

Concerned grown-ups asked questions that I would dodge. I took their offerings of Bible verse and story and hid them beneath my bed, to be brought out at quiet times. I kept the letters from pen pals and the stamps from foreign countries and I told the man and woman at the post office some jokes. I ambled the sidewalks, looked up at the leaves, listened to mourning doves, chatted with pals on the steps of the church, rode my bike anywhere I wanted to go, even past the edges of the town. I never passed the tavern, and I avoided the stares from that weird (some said retarded) man who lived alone and spent his days on his porch. I played jacks, petted dogs, swapped notes. I learned to say dirty words, even though I had no idea what they meant. My boyfriend, Toby, gave me a gumball machine ring, a rectangular green stone set in bendable gold.

The town was heaven. Not that it was really that way. All is comparison and perspective—what I chose to see, what

I didn't have to know. I didn't know about unemployment and pollution and the other weights that pushed under the people. I didn't care about commodity prices and the sale of beans and hogs. The world was sun and dark, and I would choose my own gradations and ignore the rest.

Buffalo, this town where we lived, was right down the road from Borden Chemical.

Borden and Buffalo

When Mom got sick, I decided to dredge up the dirt about Borden. She didn't link the floating fish to herself, but as her thoughts grew foggy, and she grew thinner, she kept coming back to the fish kill. She was sure Borden was the murderer. Was her body telling her something? I owed her some cause-and-effect explanation. I read books, tapped into environmental and scientific list serves, and watched documentaries. Whatever I could dig up about the chemical industry, I dug. Pesticides, herbicides, industrial chemicals, even cow shit: I know it all. Factories, coal tar, farms, sewage: put them all together, and beings die.

But Mom wanted to know about Borden Chemical. That bulbous monstrosity on Route 36. In looking there, I came across Buffalo again. I came across people. Those who I knew, whose fates were tied to the wind that blew vinyl chloride across their town. Why do we bring disaster upon ourselves? Embrace it, fund it, serve it? Facts don't mean that much. It's feeling that matters. It's earth, the ground where we plant our feet. It's what we do to compromise, what we ignore in order to stay in one place. To understand Borden, I'd have to understand the town that allowed the factory to stay: Illiopolis. And to understand Illiopolis, I'd have to go back to the town where I lived for two years: Buffalo. These were the towns nearest to Borden. I lived there. I drank that water and breathed the air. I had cancer, my mother and uncle died of cancer. Is this what fed the cells? Made us dizzy, made our joints creak, created multiplication inside that we couldn't even feel until it was late and it had compromised our systems? In comprehending these places, I hoped to get a grip on death and rage and the dark side and sex and all that stuff that comes up in nightmares, as if by grasping it all with logic I could keep from being consumed. Maybe someone would stop it.

Nobody cares about rural people, though. Let's be real about that. They're the butt of jokes; they have no power. A friend who teaches in Illiopolis joked that everyone in the town is inbred. They all have the same last names. Uh-huh, I said. Tell me about it. Feuding strands of my cousins' family, the Pattons, wind all over Central Illinois. The inbreeding isn't a matter of genetics, but of attitudes and ideas and career options and life choices that limit them to a 10-mile radius. Kids look to the chemical plant, figuring they can get a job there after they get married. They marry soon after they graduate high school, or even before. Some want to leave but lack opportunity or nerve, so they stay stoned, vandalize, fight, or maybe kill themselves driving too fast on the curving country roads.

All small towns are turned inward upon themselves: behind the times, idyllic, poor, happy, and at the mercy of whatever industry runs the place. Right at the literal center of Illinois, Illiopolis cropped up in 1833, when pioneers cleared the prairie grass and planted cabins. A fire wiped them out, leaving the land to weeds, until the railroad line from Springfield to Decatur made the town a loading point. The prairie was drained and there it appeared: farmland. Fed by decomposing roots, rich because for centuries it had been let be. Illiopolis grew along the tracks; by 1900 it had a railroad depot, grain elevators, a post office, a grocery store, a hardware store-mortuary combination (making it easy to hammer down the coffins), a lumber yard, blacksmith shops, two hotels, three churches, livery stables, a school, and, of course, a bar. Not a whole lot is different today, except that the shops that once serviced horses now service cars.

A white church with an elegant steeple looms over frame houses with added-on bedrooms. A ranch house huddles next to a rehabbed mansion, a trailer rests perpendicular to a beauty shop, a junkyard lurks up the hill from thin-walled family homes. Its bar proclaims Habits and Vices, a cavalier admission, a plainspoken truth. Johnson's Grocery sits beside the Citgo gas station with its all-night stock of liquor, chips and cigarettes. In the Business District—you know it's the business district because that's what the signs tell you—is the bank and an antiques store crammed floor to ceiling with ceramic figurines and Depression glass. That's it. Kids still drive around and around this "square" at night, swigging from beer cans in cup coolers. They honk at each other, swap joints, and screw in back seats. Then they settle down and get to the grinding work of factory shifts and child rearing. And when their children grow up, those kids drive around the square...

Relics Remain

After my parents died, I drove a couple times a week down Route 36, straight through Buffalo. It became familiar, like a rusted Chevy with bad sparkplugs. The change came not in design, but in wear. Like my parents, the town had gotten old, and it seemed to be on the verge of dying, too.

The gray one-story befits my father's grimy years of commuting to Decatur's Firestone plant. I want to hate my former home. I want to understand my mother's furious depression. But although I can see that the house is ugly, it doesn't seem ugly to me. Lilies of the valley once drooped in the shade at the side of the house, peony bushes and rhubarb grew beside the storm cellar, the laundry line ran from the house to the shed, and an alleyway provided a path where my friends and I walked barefoot. I imagine that these relics still remain. The old neighbor's tended lawn, the tree that I climbed in the front yard, the holes that my brother and I dug with Tonka trucks. What kids remember.

Route 36 used to be the only road that linked the cities of Springfield and Decatur. This pitted two-lane takes you through or near a string of towns: Illiopolis, Lanesville, Mechanicsburg, Dawson, Buffalo, Riverton. As a child, I knew the road mostly as one that shouldn't be crossed. The countryside around it is littered with faded cafes and junked VW vans, silos and leaning barns. Cattails and coreopsis droop in ditches, while stands of trees clump amidst nothing. Make-out roads stop at a field's edge or a washed-away bridge. Folks live in the middle of nowhere because it's cheap and they're comfortable; as my mom said, "We don't have to put on airs. I can go out without a bra on and I'm invisible," or, according to Grandpa, "I can scratch my ass and nobody's going to take a picture." Along the road you'll see a house grown over with vines and weeds, whatever's visible needing a paint job, greeting visitors with a No Trespassing sign and a pit bull/hound. People keep a good distance.

Near the chemical plant, the view evolves from corn to boxcar. Abandoned train cars line the right side of the road, though the active tracks are on the left. On one car, someone has spray-painted this profound truth: Decatur Sucks. Spew can be seen in the sky miles before I see the factory. Bunkers and dilapidated buildings left from the munitions grounds announce the "industrial complex." Fencing higher than my head surrounds an array of pipes and wire and bulbous tanks and fences and stacks. KEEP OUT! PRIVATE! DANGER! warn the signs. As if anyone enters by choice. These jobs are necessities.

Borden's specialties were resins and formaldehyde. The Illiopolis plant once produced my childhood friend, Elmer's Glue, and a competitor to Saran Wrap called Resinite. Borden didn't start out being one of the largest producers of plastics in the world. If you're old enough, you may remember Elsie the Cow, the smiling mascot of Borden's Dairy. When I was a six year old, living in the desert, Elsie represented everything good about the Midwest: a place of endless green, plentiful food, and kind relatives. I even had an aunt named Elsie. So imagine my sense of betrayal when I realized that Borden had gone from producing milk to producing chemicals. While it's hard to imagine doing without plastic wrap, the fact is that it leaches into foods when microwaved. The white glue that we used in school is made with non-toxic levels of the same chemicals. The big product in Illiopolis now is polyvinyl chloride (PVC), used in tile, plastic water pipes, siding and wire insulation. What goes into PVC? Vinyl chloride, a flammable gas, and vinyl acetate, a toxic gas. When they shoot into the air or get into the water, they cause everything from a bad cough to paralysis. Workers in plants that make the stuff have high rates of liver and breast cancer. Vinyl chloride stays in water for decades, where it is absorbed into fish flesh. The fish we ate on those glorious summer cookouts.

Borden Chemical has undergone some particularly colorful corporate transformations. In 1987, according to a spokesperson, it came under the ownership of Borden Chemicals and Plastics Limited Partnership (BCP), which is now twisting in the wind of Chapter 11 bankruptcy. In April of this year, the Illiopolis plant was sold to Taiwan-based PVC-maker Formosa Plastics, which not only has a record of environmental violations in several states but was also tied to campaign finance scandals in the Clinton White House.

Chemicals were and still are dumped in rivers and wells and underground tanks and landfills, floating and settling all across the Sangamon Valley. Companies dodge environmental protection laws by getting waivers and exemptions. Since we need our conveniences, the wastes have to go somewhere—why not some hick's backyard?

When the fish turned up dead, my relatives blamed Borden. But with so many companies and farms polluting the Sangamon River, we can never be sure. But Borden has an unenviable environmental record. It was even caught shipping 2,500 drums of highly toxic mercury waste to South Africa. The stockpiled drums leaked contaminants—a disaster that led to both criminal and civil investigations in South Africa.

At one of Borden's biggest plants, in Geismar, Louisiana (home of BCP), according to Time magazine, a "witches' brew of toxic chemicals" descended not once, but several times. The chemicals that Borden reportedly released into the air were ethylene dichloride, vinyl-chloride monomer, hydrogen chloride, hydrochloric acid and ammonia. In 1994, three years before Geismar's first toxic spew, the U.S. Justice Department filed a lawsuit accusing the company of illegal hazardous waste storage, contaminating groundwater, burning waste without a permit and neglecting to report chemical releases into the air. In 1998, without admitting wrongdoing, Borden settled for $3.6 million and agreed to spend $3 million to clean up the Louisiana water.

The Illinois Pollution Control Board exempted Illiopolis' Borden plant from many Environmental Protection Agency

(EPA) regulations. Borden's wastewater is still high in "total dissolved solids, or TDS"—all of the solid contaminants put together, like calcium, magnesium, iron, lead, nitrates, chloride and sulfate. This is twice as high as the average, and the Illinois Department of Energy and Natural Resources says that these flows are carried downstream to the Sangamon River. The plant can legally dump 800,000 gallons of wastewater into the stream every day. In two miles, this water reaches the Sangamon. Because the area around the stream is wooded, locals hunt and fish there. EPA studies show the river from Decatur to Springfield to be especially toxic: the river where we caught fish, used water for crops, and walked when the currents were low. The river that one naturalist calls "a drainage ditch." Because the water runs low and bends, poisons gather, plants die, and invasive species bloom. Little is washed away.

In May of 1999, Borden Chemicals and Plastics did to Illiopolis what it did to Geismar, Louisiana: It released 500 pounds of vinyl chloride gas. An "accidental" release, they said. Ten years before this, in 1989, the Illinois Attorney General's Office sued the plant for releasing the same gas 14 times over a four-year period. In response to the 1999 incident, Borden was forced to hold public hearings detailing an emergency evacuation plan. Not counting accidents, the plant routinely releases 65,000 pounds of vinyl chloride and 40,000 pounds of vinyl acetate into the Central Illinois air every year. Along with causing cancer, vinyl chloride is suspected of disrupting our hormones, which makes us—especially women and girls—vulnerable to all kinds of illnesses. I was not reassured by what I found out about Borden Chemical and its history of environmental abuses, but I can't assume the same about the residents of Illiopolis.

Why did the thousand people in this town welcome Borden Chemical, working there, sending their children to work there? Because they need to eat. And because they can't do anything about it, and around here, people don't worry about what they can't fix. What are they going to do, drive to Decatur to work at Archer Daniels Midland, where four employees died in an assortment of industrial accidents? Or to A.E. Staley, with its 12-hour shifts? At least it's a job.

Blue-Collar Risks

When I visited the family over Easter, Uncle Wade talked about men who worked in a Taylorville factory. Each had died or was dying from a rare cancer. They'd been forced to climb into tanks to clean out mystery chemical crap. Anyone who said no was fired. "They swelled up big as blimps before they kicked off," Wade said. "Pathetic bastards."

Since he owned his own construction business, my uncle never had to worry about being in any employer's pocket. They took on other blue-collar risks: My uncle can barely move from motion injuries. But they chose the work, and that makes all the difference. My cousin Mike, who took over the business from his father, says if you don't become "a boss" by 40, you'll turn into a cripple. "Your joints stop working," he said. He told stories of danger and carelessness: men with hands stapled to walls, men with nails driven into legs. Which tools cause arthritis in the long run. "You got to know what you're doing." But, he said, it's better than being a factory cog.

We had thought our choices made us safe. We lived in the country by the river, away from the city and its violence. We never traveled, never took a plane or train or bus, never drove farther than "town." My family never went to college, where they might meet scary people. But for all that, my mother spent the Christmas of 2000 stuck in a chair because of the swelling in her legs. The tumor caused the swelling, she claimed. It was only later, by stealing a look into her medical chart that I found out her new diagnosis: lymphoma.

"So, did you ever find out anything about that fish kill?" I asked Mom, as we nibbled Teresa's peanut butter fudge and watched through the windows at the kids sledding down the long, steep riverbank. Mom didn't look sick—more lined, puffier, but normal enough. I bet then that she might have a few years left. She never mentioned time, never talked about her illness, and if pressed, she lied. I considered calling her doctor for the truth, but what would be the point? We knew it was beyond surgery, and that chemo wasn't going to do it. Mom was still convinced that the treatments would work, though, and blamed everything on her doctor. "She's given up on me," was her line. "She thinks I should go into a hospice. She wants to shove me into a corner so she doesn't have to deal with me." Chemo splints and radiation blasts were getting it, Mom insisted, even though the doctor said otherwise. Though Mom grew more bloated and forgetful, though she rarely ate. It was easy to take her word for it, as if what I saw was just a delusion, and the will truly was greater than the body. Families can convince themselves that what we choose to believe is what is true.

"Borden might've done it," I said. "But other factories were doing the same thing."

"It was Borden," she insisted. "That's what your grandpa said."

"How did he know?"

"It was in the paper." But I looked for proof in the newspapers, and never found it. There had been other fish kills that year. The signs pointed more to the Decatur Sanitary District than Borden. "It could have been anyone," I told her.

"Everyone knows it was Borden. You'd think they would've had more respect for the fish."

"What about respect for the people?"

"Well, nobody ate the fish after that. Nobody died," she said, lighting up a cigarette. ❖

Reprinted with permission from **E/The Environment Magazine** Subscription Department: P.O. Box 2047, Marion, OH 43306 Telephone: (815) 734-1242 (Subscriptions are $20 per year) On the Internet: www.emagazine.com email: info@emagazine.com

Questions

1. What is polyvinyl chloride (PVC) used in?
2. What chemicals did one of Borden's biggest plants, in Geismar, Louisiana, reportedly release into the air?
3. How many gallons of wastewater can Illiopolis' Borden plant legally dump into the stream every day?

Answers are at the back of the book.

In one of the most glaring examples of corporate irresponsibility, RCA is being sued for almost $7 million in compensatory damages by ill Taiwanese residents. Former workers are claiming that exposure to solvents at the old RCA factory have led to higher cancer rates end environmental contamination. For over two decades, workers were regularly exposed to toxins without warning of their effects, access to protective gear, or proper ventilation. This case has brought to light the hazards hidden in hi-tech industrial production.

Facing Up to a Dirty Secret

Erling Hoh

Far Eastern Economic Review, December 12, 2002

"BUILDING TRUST THROUGH QUALITY." The old RCA slogan still hangs on a wall inside the company's Taoyuan factory, but the gleaming machines that ushered in Taiwans hi-tech revolution, and the 10,000 workers who operated them, are long gone.

Three decades ago, young workers from all over the island flocked here in search of jobs. Huang Tzu-ping, who was 17 at the time, remembers the clean, beautiful factory. "It had the best environment in the electronics business," she says.

Today, all that's left are dilapidated, cavernous factory buildings and a stretch of severely poisoned land.

"This is one of the most serious contamination sites and one of the worst cases of global corporate irresponsibility that has been uncovered around the world," says Ted Smith, executive director of the Silicon Valley Toxics Coalition, a citizens' group that monitors the hi-tech industry's impact on the environment and on human health.

Between 1970 and 1992, toxic organic solvents used in the production of electronics equipment were systematically and illegally dumped on the factory grounds. When Taiwan's Environmental Protection Agency (EPA) conducted groundwater tests in 1994, two years after the RCA plant closed, it found that the levels of toxic solvents in the water were up to 960 times higher than those considered safe for human use. Yet for more than two decades, workers in the factory and inhabitants in the neighbourhood drank and used this water.

Former workers at RCA claim exposure to the solvents led to unnaturally high rates of cancer. A group of 263 are now seeking $6.6 million in compensation for illnesses they say were brought on by working at Taoyuan. The company that eventually bought the RCA division at Taoyuan, French state conglomerate Thomson Multimedia, has publicly denied any link between the use of solvents and cancer among workers.

For Taiwan—which President Chen Shui-bian wants to see become a "green, silicon island"—the revelations of what happened at Taoyuan have served as a wake-up call to the environmental threat from the electronics industry. Dumping of the kind carried out in Taoyuan is now treated as a criminal offence with a maximum penalty of life in prison. Since 2000, the government has had the retroactive power to force polluters to clean up affected groundwater, or pay fines of up to $140,000 a day.

"This is Taiwan's Love Canal," says Dr. Harvey Houng, an EPA adviser, referring to the toxic landfill disaster in upstate New York that put the environment on the United States' national agenda in the late 1970s. "The hi-tech industry is the largest generator of hazardous waste in Taiwan."

That may surprise some, but it shouldn't, says Dr. Joseph LaDou, a professor of occupational and environmental medicine at the University of California-San Francisco: "The public perception is that this is a clean industry," he says. "It's only clean if you are a chip."

RCA—the Radio Corporation of America—first came to Taiwan in the late 1960s, when it moved production of its black-and-white television sets from Memphis, Tennessee, to Taoyuan, a town about 30 kilometres west of Taipei. That huge investment in the plant, which also produced computer components, was a crucial, first step in Taiwans development into a hi-tech economy.

"It became one of the first Fortune 500 companies to move overseas," says Jefferson Cowie, author of Capital Moves: RCA's Seventy-Year Quest for Cheap Labour and an assistant professor at Cornell University's School of Industrial and Labour Relations. "They looked for the same things companies look for today: high employment, low wages and an abundance of young women." (Paradoxically, environmentalists now accuse Taiwanese companies of following RCA's example in less developed parts of Asia.)

Like other companies in the hi-tech sector, RCA used organic solvents to dean and degrease components. Huang Tzu-hui, an RCA worker for io years, recalls dipping printed circuit boards into the solvents without wearing gloves. She also says the used solvents were poured into toilets, sewers, or simply thrown onto the grass outside.

Chang Shu-mei, who worked at the RCA plant for 14 years, recalls how some young girls vomited during their first days on the production line in the power-cord department. Others, who could not get used to the noxious fumes from the melted plastic pellets, quit after only a few days. "Sparrows that had strayed into the factory became so dazed that we could pick them with our hands," says Liang Ko-ping, who worked in the factory for 18 years, and who now heads the RCA Employee's Association.

Liang and other workers say they were not informed of the solvents' toxicity, and did not receive proper instructions on how to use and dispose of them. Taiwan's Council of Labour Affairs supports these assertions: Between 1975 and 1991, the government agency inspected the RCA Taoyuan plant eight times. Each time, it found the company in breach of safety rules governing the use of organic solvents: Ventilation was insufficient, solvents were not labeled, advice on the effects of solvents on the human body was not posted, workers were not informed how to handle emergency situa-

tions, routine and legally required health checks were not carried out.

In 1986, RCA was purchased by another U.S. electronics giant, General Electric. A year later, the French state conglomerate Thomson Multimedia in turn purchased GE's consumer-electronics division, which included the RCA operations at Taoyuan. In 1989, Thomson surveyed the groundwater at the plant and became aware of the magnitude of the problem. According to a doctoral dissertation by Mong Weider, a professor at the Police University in Taoyuan, Thomson initially planned to announce the findings at a press conference, even rehearsing for it, before deciding not to make public the severity of the situation. Thomson has declined to comment on this assertion.

In 1992, the factory was finally shut down and the site was sold to a local developer. Two years later, the EPA conducted its survey of the groundwater in the area and found that it was seriously contaminated with organic solvents. GE and Thomson immediately agreed to supply bottled water to people living in the area—who number about 4,000 today—until water from another source could be piped in, and reached an agreement with Taiwan's EPA to decontaminate the area. The work began in 1996 and ended in 1998.

"The decreasing level of chemicals in the groundwater show that the action undertaken by GE and Thomson in 1996-98—removing and treating over 10,000 cubic yards of soil—is working," says Gary Sheffer, a GE spokesman. Surveys show the soil pollution has improved, but groundwater contamination remains severe, and the local government has asked RCA-Taiwan to submit a pollution-control plan for the area.

■ ■ ■

The problems at Taoyuan are not unique: Organic solvents have always been a ubiquitous part of the electronics industry. While debate continues over the nature of the health risk they pose, their status as pollutants is well established. California's Santa Clara County, the home of Silicon Valley, for instance, has 23 sites on America's EPA National Priority List of the country's most polluted locations—more than any other county in the U.S. Eighteen of these sites involve groundwater contamination by organic solvents from the microelectronics industry.

In March this year, the Semiconductor Industry Association, a U.S. industry grouping, announced it would conduct a preliminary review to determine if it is possible to conduct a study of health risks in the industry. But many medical researchers believe there's already cause for concern.

The International Association of Research on Cancer, a World Health Organization agency, classifies tetrachloroethylene and trichloroethylene, two of the solvents that have contaminated the groundwater in Taoyuan, as possible human carcinogens. Meanwhile, in one of the most widely publicized studies in recent years, female workers at the National Semiconductor plant in Greenock, Scotland, were found to have stomach cancer rates four to five times higher than expected, and lung cancer rates two to three times above the mean.

In the past few years, several epidemiological studies have been conducted to establish whether or not the RCA workers and inhabitants in the area are suffering from unusually high rates of cancer and other diseases. While not willing directly to link the use of the solvents used at RCA with cancer, Dr. Hwang Jung-der of National Taiwan University, who conducted a health study of residents near the RCA plant, says some of the contaminants used at Taoyuan "were probable human carcinogens."

Another investigation by the Council of Labour Affairs showed only marginally higher rates of breast cancer among female workers at the plant. Critics, however, contend that a scientifically accurate survey is impossible without access to the Taoyuan plant's employee records. Thomson says it does not have these records, and believes they were destroyed in a fire at the plant in 1985.

But former RCA workers and nearby residents say it is not a question of science and statistics, but of personal loss. Three guards who worked at a post near where the solvents were dumped have died of cancer. Nearby the factory, there is a street known as "Widow's Alley" along which ii residents have died of cancer in recent years, according to neighbourhood council chairman Huang Chin-chiang. One worker, Chang Shu-mei, who spent seven years at RCA and was subsequently diagnosed with breast cancer recalls: "I didn't vomit, I was just dizzy." She adds: "Three of my co-workers also had breast cancer."

It's likely to be years before the RCA workers claims are finally resolved. But regardless of the outcome, the workers believe it's vital that Taiwan learns a lesson from what happened at Taoyuan. "Taiwan is hi-tech. Everybody knows it is the future of Taiwan," says Wu Chi-kang, who spent 18 years at the RCA Taoyuan plant. "But what people don't realize are the hazards of this production. We want to make people aware of these dangers." ❖

Questions

1. What did Taiwan's Environmental Protection Agency find when it conducted groundwater test in 1994, two years after the RCA plant closed?
2. How was the RCA Taoyuan plant in breach of safety rules governing the use of organic solvents?
3. What county in the United States has the most sites on America's EPA National Priority List of the country's most polluted locations?

Answers are at the back of the book.

Millions of citizens of rural Bangladesh are faced with the danger of arsenic poisoning from drinking water. Without a monetarily viable way of testing millions of wells throughout the country, thousands face a future of skin lesions or cancer. Recently, however, a relatively inexpensive device that can test or filter drinking water for contaminants has been in development. The solution may be on the horizon.

18

Attacking an Arsenic Plague

Helen Epstein

Popular Science, November 2002

FOR MILLIONS OF PEOPLE IN RURAL BANGLADESH, drinking water is like playing Russian roulette. Thousands will die in years to come from cancer caused by arsenic, a natural element in their water. Thousands more will suffer hideous skin lesions. There is no simple, cheap The machine uses a traditional way to test drinking water for contamination. The Bangladesh government—recently named the world's most corrupt by a widely respected international monitoring agency—has received $32.4 million from the World Bank to determine which wells are safe, but is hampered both by its notorious bureaucracies and the sheer scale of the problem: Some 10 million private wells need testing. Existing arsenic-measuring devices are expensive and awkward; some also produce a toxic arsene gas. But Pietro Perona, a Caltech electrical engineer, believes he has a solution. It's called the arsonometer, a gizmo about the size of a Walkman, rigged up from a pair of infrared LEDs, two glass tubes, a couple of photodiode detectors, an LCD display, and a 9-volt battery.

The machine uses a traditional method for arsenic detection, in which molybdate (a salt derived from the element molybdenum) is added to two samples of groundwater. In the presence of oxidized arsenic, molybdate turns water blue. The

problem is, molybdate also turns water blue in the presence of phosphate, which is abundant in water. So before molybdate goes in, a reducing agent is added to one water sample and an oxidizing agent to another. Since molybdate doesn't react with reduced arsenic this sample will be lighter than the oxidized sample. The concentration of arsenic can be determined by measuring the difference in the quality of blue between the two samples. This is accomplished by the LEDs, which send light through the glass tubes, and the photodiode detectors, which measure the amount of light in each sample. The arsenic concentration is quantified from light readings. (Currently, the device can only be used by trained technicians, but it could easily be redesigned to display results in parts per billion, so virtually anyone could use it.) Perona's collaborator, Alexander van Green of Columbia University is currently in Bangladesh, testing the device, which would cost $50.

Meanwhile, Fakhurl Islam, a chemist at Bangladesh's Rahshahi University, has invented a cheap filter that strips arsenic from drinking water. Arsenic binds to ferrous sulfate, a cheap, blue-green crystalline powder. Islam took dust fro the simple mud bricks used throughout the country, soaked it with ferrous sulfate solution, and then baked the concoction at a high temperature. A 20-kilogram bag of this mixture costs about $3 and can be used in filters to purify 3,000 liters of water. After the filter is exhausted, the arsenic-saturated brick dust can be thrown away—the arsenic won't come off.

"We're really excited about it," says Courtney Bickert of International Development Enterprises, a nonprofit organization that supported Islam's research. "It's locally invented and produced, field tests show that it works, and there is no hazardous waste site."

Islam named his filter Shapla, after his country's national flower. Some observers, such as Colin Davis, a UNICEF water expert, hope that constructing filters and arsonometers might become a cottage industry in Bangladesh. If the government has been criticized for failing to solve the arsenic crisis, says Davis, "the private sector is very vibrant in Bangladesh. Once the Shapla filters and arsenometers become widely known, the solution could be found in private enterprise." With more than a little help from science. ❖

Questions

1. How many private wells in Bangladesh need to be tested for arsenic?
2. What is the arsenometer?
3. What did Fakhrul Islam, a chemist at Bangladesh's Rahshahi University, invent?

Answers are at the back of the book.

19

Although efforts to reduce the emission of chemicals that cause acid rain have been underway for decades, a new study suggests that such efforts may not be enough to prevent long-term damage to the environment. Acid-causing substances have been building in the soil over many years, which could cause chemical effects for decades to come. Given the complexity of the effects of acid rain, a call for more comprehensive pollution controls is being sounded.

Long-Term Data Show Lingering Effects from Acid Rain

Kevin Krajick

Science, April 13, 2001

ACID RAIN IS LIKE THE PROVERBIAL BAD PENNY: Every time you think you've passed it off, it shows up in your change again. Studies over the past few years have turned up more and more evidence of its lingering effects. Now a major research synthesis provides the most comprehensive view to date—it is not encouraging—and prescribes a drastic cure. The call for more stringent controls comes just in time to fuel what promises to be a ferocious debate over stricter federal regulation of not only acid emissions but also the main greenhouse gas, carbon dioxide.

Progressively tougher pollution rules over the past 3 decades have reduced U.S. emissions of the main acid rain ingredient, sulfur dioxide (SO_2) by about 40% from its 1973 peak of 28.8 metric tons a year. By 2010, SO_2 emissions should be less than half of the 1973 levels. But in the March issue of *BioScience*, 10 leading acid rain researchers say victory toasts are premature. They say power plants, the main contributor, must cut SO_2 emissions another 80% beyond the current mandate to undo past insults to sensitive soils and waters in the northeastern United Sates and, by implication, elsewhere. They also assert that these reductions, which would amount to an overall 44% cut in sulfur emissions, may bring only

partial recovery to fish and trees by 2050. At the same time, acidifying emissions of nitrogen oxides (NO_x)—still relatively less regulated—are level or even mounting, causing collateral damage.

"Coming from such a consensus, this study solidifies many things," says Rona Birnbaum, chief of the Environmental Protection Agency's (EPA's) acid rain assessment program. "There was uncertainty especially over long-term soil impacts. Now it's undeniable." James Galloway, chair of environmental science at the University of Virginia in Charlottesville, adds that "the old controls were clearly not enough. Acid rain is a lot more complex than we first thought."

The problem is that acid-causing substances have built up in the ground and are still causing cascading chemical effects that could continue for decades. The first good evidence of acid rain's long-term effects came in 1996 from New Hampshire's Hubbard Brook Experimental Forest, where data have been collected since the early 1960s. That study showed that half the base cations of the nutrients calcium and magnesium, which neutralize acids, had leached from soils in the past few decades; as a result, vegetative growth was near a standstill (*Science*, 12 April 1996, p. 244). The researchers blamed excess acids in the soil for dissolving the cations into drainage waters much faster than weathering bedrock below could replenish them. Adding to the insult, they said, was the fact that smokestack scrubbers installed to reduce particulates were also removing soot rich in calcium that had previously replaced some nutrients the acids were leaching.

The new study, a collation of further Hubbard Brook data and recently published studies from many other sites, adds considerable breadth and detail. Gregory Lawrence, a research hydrologist with the U.S. Geological Survey (USGS) and a co-author, says many northeastern soils "have a long memory: for acid deposition. Residual sulfur and, to a lesser degree, nitrogen are being slowly released by microbes and plants, he says, bumping the cations off the soil particles they normally cling to. This makes the cations highly soluble and easily washed away during rains and snowmelts. As a result, nutrient levels at many sites are showing little improvement, and some are actually worsening. Some of the hardest hit areas overlie calcium-poor sandstones in the Catskills mountains in New York state, where nearly all nutrients have disappeared in places, right down to glacial till and bedrock. Work by Lawrence and others also shows that once nutrients are depleted, excess acids mobilize the soils' abundant aluminum; usually held in harmless organic form, aluminum is poisonous when it dissolves.

Effects on aquatic life have been known since the early 1990s. Some 15% of lakes in New England and 41% in New York's Adirondack Mountains are chronically or episodically acid, says the report, and many such lakes have few or no fish.

Good evidence of effects on trees has been slower in coming, but that too has mounted lately. Although the exact role of acid deposition is unclear, extensive die backs and loss of vigor among red spruces and, more recently, sugar maples have been chronicled for at least 10 years. Co-author Christopher Cronan, a biologist at the University of Maine, Orono, says maples especially in Pennsylvania appear to be malnourished; at some sites, aluminum is glomming onto rootlets, blocking uptake of whatever meager nutrients are left. In spruces the problem is compounded. In 1999 researchers at the University of Vermont in Burlington traced much of their decline to baths of acid fog and rain that leach calcium directly from needles—a reaction that leaves needle membranes unable to cope with winter freezing. It's death by a thousand cuts, says Cronan: Instead of killing directly, acid rain usually leaves trees susceptible to drought or insects, which finish the job. Weakened conifers in southern Appalachia, for instance, are now being defoliated en masse by invading woolly adelgids, exotic parasitic insects. The

chain reaction continues. The endangered spruce-fir moss spider, a small tarantula that needs shades from the trees, is now found at just a few known sites, according to entomologists.

This is just one signal that the problem is not limited to the Northeast. Arthur Bulger, a fish ecologist at the University of Virginia in Charlottesville and also a co-author, says that acid rain effects are now becoming more apparent in the Southeast, some 20 years after they appeared in the Northeast. He and colleagues have chronicled these effects in a paper in the *Canadian Journal of Fisheries and Aquatic Sciences* last year and another in press at *Hydrology and Earth System Sciences*. The reason for the time delay, says Bulger, is that southern soils are generally thicker than northern ones and thus able to sponge up far more acid before leaking it to the surrounding environment. But now that soils are saturated, acid levels in nearby waters are skyrocketing. Bulger's colleagues have studied 50,000 kilometers of streams; in a third, he says fish are declining or are already gone. He predicts another 8000 kilometers will be affected in coming decades.

Until recently, acid rain has been largely off the radar screen in the western states, in part because the population is smaller and the coal that fuels power plants there is much lower in sulfur. But now the region's cattle feedlots are booming, as is the human population. The former churn out lots of manure, and the latter insists on driving more and larger motor vehicles. Both produce acid-causing NO_x. Jill Baron, a USGS ecologist in Fort Collins, Colorado, and colleagues are studying 300- to 700-year-old spruces in the Rockies. In the journal *Ecosystems* last fall, they said trees downwind of populous areas show high levels of nitrogen and low ratios of magnesium in their needles. Nearby streams are also showing dramatic changes in their populations of diatoms, shifting from species that do well in the region's usually nutrient poor waters to those common in overfertilized waters. "Too bad most people get a lot less excited about diatoms than about fish," says Baron. "We're not nearly as bad as the East, but we're beginning a trajectory that will take us there."

The Northeast research team has assembled its data into a model to project the possible impact of various emissions cuts. Lead author Charles Driscoll, director of Syracuse University's Center for Environmental Systems Engineering, says that at current regulatory levels, the most sensitive environments such as Hubbard Brook will probably cleanse themselves, but very slowly. Chemical balances might return by about 2060, he says; after that, lake zooplankton might come back with 10 years; some fish populations, 5 or 10 years after that. If Congress were to call for an 80% SO_2 reduction from power plants below the current target for 2010, streams would probably bounce back by 2025, and some biological recovery in them might come by 2050.

"The question of soil and trees is a lot harder to answer," says Driscoll. In a communication to *Nature* last October, John Stoddard, an EPA scientist in Corvallis, Oregon, suggested that some soils with high sulfur-adsorbing capacities such as those in the Southeast might take centuries to recover.

The *BioScience* study received major attention on Capitol Hill, where momentum for stricter air controls has been building. Just before the study came out, President George W. Bush announced that he was breaking his campaign promise to limit CO_2 emissions—produced largely by the same power plants implicated in acid rain—raising the political stakes in the debate. With the 1990 Clean Air Act amendments up for reauthorization this spring, a half-dozen bills have been knocking around, proposing 40% to 65% reductions in both SO_2 and NO_x. On March 15, New York Representative Sherwood Boehlert and Vermont Senator James Jeffords, both Republicans, introduced companion proposals, calling for a

75% cut in SO_2 below what is currently mandated and a 75% cut in NOx form recent levels—and a rollback of CO_2 to 1990 levels. Boehlert, head of the House Committee on Science, said the *BioScience* paper "is a wake-up call, and it should lead anyone who truly believes in a science-based policy to support acid rain control."

Bush has said he will support stricter acid controls, although he hasn't mentioned any numbers. Moderate Republicans, who are furious that he reneged on CO_2, vow to keep all the pollutants tied together—a strategy that could greatly complicate a solution. Dan Riedinger, a spokesperson for the power industry's Edison Electric Institute, says the call for huge reductions now "is a little premature." He points out that ozone-curbing regulations scheduled to start in 2004 will also cut acid emissions and that some controls mandated in 1990 kicked in only last year. "The current program needs to be given more time to work," he says. "We always knew it would take decades." That last part may be the only thing on which everyone agrees. ❖

Questions

1. Progressively tougher pollution rules over the past 3 decades have reduced U. S. emissions of sulfur dioxide by how much?
2. What percentages of lakes in New England and in New York's Adirondack Mountains are chronically or episodically acidic and how does that affect aquatic life?

3. If Congress were to call for an 80% SO_2 reduction from power plants below the current target for 2010, how soon would streams bounce back and some biological recovery in them come?

Answers are at the back of the book.

20

Because of massive international efforts to reduce emissions of certain chemicals, the hole in the ozone layer is mending itself. At a time when environmental issues are so varied and complex, this good news can serve to show what can be accomplished when global communities unite and work together towards solutions.

News on the Environment Isn't Always Bad

Mark Sappenfield

The Christian Science Monitor, October 4, 2002

IN THE WORLD OF ENVIRONMENTALISM, things can often seem rather bleak. Rare species from Barrow to Borneo are likely going extinct before they are even discovered. Rain forests are shrinking. Greenhouse gases, it sometimes seems, are turning the atmosphere into a giant toaster oven.

Then came the news late last month: According to Australian scientists, the hole in the ozone layer—a symbol of human environmental destruction so universal that it became the punch line in an Austin Powers movie—will begin closing in three years. Thanks to international efforts to ban certain chemicals, the opening would shut by 2050.

In light of the sense of approaching apocalypse on many conservation issues, this is a success story more common than many people might expect. It highlights a history of progress on some of the most serious environmental problems of the past 30 years—from clean air to panda bears.

The progress is far from complete. Environmental threats are perhaps more varied and widespread than they have ever been. Yet the success in the atmosphere above Antarctica, and improvements in several other areas, suggests that when the global community identifies a problem and unifies behind a solution, it can reverse even the most dire environmental disasters.

The success of some international environmental measures often receives little attention. Indeed, the ozone reforms could provide at least a framework for how to move forward on the first great green issue of the 21st century: climate change.

"It's an untold story," says Daniel Esty, a professor of environmental law at Yale University in New Haven, Conn. "It's a good example that well-designed programs can work."

Closing the Ozone Gap

Strictly speaking, the news that the ozone layer is on the mend is not a surprise. The timeline fits nicely with the one laid out by the Montreal Protocol, which in 1987 required nations to cut their use of chorofluorocarbons (CFCs), the chemicals causing the problem.

In an unrelated but confusing development this week, American scientists say the ozone hole is currently 88 percent smaller than normal. But that's a function of unusual weather, and scientists expect the hole to expand again.

By contrast, the gradual and human-induced shrinking of the hole reported in the Australian study is permanent, and in that respect, more significant. It is further confirmation of a surprisingly positive global track record on the environment—one that is often lost amid the desire to point out how bad things are, rather than chronicle how much better they have gotten in recent decades.

Environmental Improvement

As an example of the nationwide improvement in air quality, experts note that Los Angeles this year enjoyed its third consecutive summer without a single smog alert. (In the 1970s, 80 alerts a year were not unusual.) They suggest that the international campaign to save the whales saved the whales, that global bans on chemicals like DDT revitalized bird populations like the bald eagle, and that acid rain has been cut by more than half in Europe and the United States during the past few decades.

Activists will gladly—and rightly—proclaim that a phalanx of other issues from biodiversity to fishing rights has either been largely ignored by the international community or mangled by less successful regulations. But many domestic and international programs have "absolutely" had a significant impact on the environment, says Karim Ahmed of the National Council on Science and the Environment in Washington.

The Montreal Protocol "is the most dramatic example," he adds, and it offers insight into how and when environmental reforms are successful.

For one, there needs to be a crisis that grabs the public's attention—and scientific proof of it, whether it's a decline in cuddly Chinese bears or a rise in lead poisonings.

When two scientists in 1974 posited that CFCs were eating away at the ozone layer, which protects the Earth from the sun's ultraviolet rays, a few companies discontinued offending aerosol cans, but the reaction was muted. In 1985, however, researchers looked at the Antarctic atmosphere and found that CFCs were thinning a massive section of the ozone layer. Suddenly, public health—always the greatest motivating factor in environmental reform—was at risk, creating a sense of urgency.

The Montreal Protocol followed, and businesses, which had estimated that the cost of phasing out CFCs would be devastating, were pressured to find economic alternatives—and did. "Once industry, science, and politics got together, it started to move quickly," says Stephen Andersen, author of "Protecting the Ozone Layer." "It turned out to be fairly painless."

To some, there are clear parallels between the emergence of the problems with the ozone layer and the current questions over global warming. If science does prove a significant link between greenhouse gases and global warming—and can show how that is hurting humans—many expect a similar sort of urgency to follow.

And Now for the Hard Part...

But many of the similarities end there. Global warming is representative of the new environmental challenges of the 21st century, says Dr. Esty of Yale. The broad reforms of the 1980s and '90s went a long way toward cleaning up the biggest polluters. Now, global warming and other issues will have to take the fight to the ordinary modern citizen.

"The scale of this issue is 500 million times more complicated," says Esty. "You're implicating every business, every family that drives an automobile." ❖

Questions

1. According to Australian scientists, in what year would the hole in the ozone layer close?
2. What was the 1987 international treaty designed to phase out CFC's?
3. Why is global warming more difficult to solve than loss of the ozone layer?

Answers are at the back of the book.

21

According to the United Nations' Intergovernmental Panel on Climate Change, increased temperatures caused by global warming could create death, pestilence, wildfires, flooding, rising sea levels, water wars, and environmental refugees. Scientists, governments, and some corporations have acknowledged global warming and climate change as being detrimentally accelerated by human activity. The good news: if meaningful, long-term, environmental action plans are formulated and implemented, many pending disasters could be curbed or eliminated. The bad news: without meaningful action, widespread degradation is to be expected, with the poorest countries bearing the brunt of the consequences.

The Weather Turns Wild

Nancy Shute, Thomas Hayden, Charles W. Petit, Rachel K. Sobel, Kevin Whitelaw, David Whitman

U.S. News & World Report, February 5, 2001

THE PEOPLE OF ATLANTA CAN BE FORGIVEN for not worrying about global warming as they shivered in the dark last January, their city crippled by a monster ice storm that hit just before the Super Bowl. So can the 15 families in Hilo, Hawaii, whose houses were washed away by the 27 inches of rain that fell in 24 hours last November. And the FBI agents who searched for evidence blown out of their downtown Fort Worth office building, which was destroyed by a tornado last March. Not to mention the baffled residents of Barrow, Alaska, who flooded the local weather office with calls on June 19, as rumbling black clouds descended—a rare Arctic thunderstorm.

But such bizarre weather could soon become more common, and the consequences far more dire, according to a United Nations scientific panel. Last week, the Intergovernmental Panel on Climate Change met in Shanghai and officially released the most definitive—and scary—report yet, declaring that global warming is not only real but manmade. The decade of the '90s was the warmest on record, and most of the rise was likely caused by the burning of oil, coal, and other fuels that release carbon dioxide, as well as other so-called greenhouse gases. What's more, future changes will be twice as severe as predicted just five years ago, the group says. Over the next 100 years, temperatures are projected to rise by 2.5 to 10.4 degrees worldwide, enough to spark floods, epidemics, and millions of "environmental refugees."

By midcentury, the chic Art Deco hotels that now line Miami's South Beach could stand waterlogged and abandoned. Malaria could be a public health threat in Vermont. Nebraska farmers could abandon their fields for lack of water. Outside the United States, the impact would be much more severe. Rising sea levels could contaminate the aquifers that supply drinking water for Caribbean islands, while entire Pacific island nations could simply disappear under the sea. Perhaps the hardest-hit country would be Bangladesh, where thousands of people already die from floods each year. Increased snowmelt in the Himalayas could combine with rising seas to make at least 10 percent of the country uninhabitable. The water level of most of Africa's largest rivers, including the Nile, could plunge, triggering widespread crop failure and idling hydroelectric plants. Higher temperatures and lower rainfall could stunt food production in Mexico and other parts of Latin America.

No more words

"The debate is over," says Peter Gleick, president of the Pacific Institute for Studies in Development, Environment, and Security, in Oakland, Calif. "No matter what we do to reduce greenhouse-gas emissions, we will not be able to avoid some impacts of climate change."

This newest global-warming forecast is backed by data from myriad satellites, weather balloons, ships at sea, and weather stations, and by immense computer models of the global climate system. As scientists have moved toward consensus on warming's inevitability, there has been growing movement to come up with realistic adaptations to blunt the expected effects. Instead of casting blame at polluting SUV drivers, environmentalists and businesses alike are working to create feasible solutions. These range from measures as complex as global carbon-dioxide-emissions taxes to ones as simple as caulking leaks in Russian and Chinese natural gas pipelines. The take-home message: Change is difficult but not impossible, and the sooner we start, the easier it will be. Civilization has adjusted to drastic weather changes in the past (box, Page 52) and is well positioned to do so again. Indeed, while governments squabble over what is to be done, major corporations such as BP Amoco and DuPont are retooling operations to reduce greenhouse gases. "I am very, very optimistic," says Robert Watson, an atmospheric scientist, World Bank official, and leader of the IPCC panel that created the report.

Concern about greenhouse gases is hardly new; as early as the 1700s, scientists were wondering whether atmospheric gases could transmit light but trap heat, much like glass in a greenhouse. By 1860, Irish physicist John Tyndall (the first man to explain why the sky is blue) suggested that ice ages follow a decrease in carbon dioxide. In 1957, Roger Revelle, a researcher at the Scripps Institution of Oceanography in California, declared that human alteration of the climate amounted to a "large-scale geophysical experiment" with potentially vast consequences.

Such dire predictions had been made before and not come true, and this environmental hysteria emboldened skeptics. But by 1988, the evidence was hard to rebut; when NASA atmospheric scientist James Hansen told a congressional hearing that global warming had arrived, climate change became a hot political topic. At the 1992 Rio de Janeiro Earth Summit, 155 nations, including the United States, signed a treaty to control greenhouse emissions, which also include other gases such as methane. That accord led to the 1997 Kyoto protocol calling for reducing emissions of developed nations below 1990 levels but placing no emissions restrictions on China and other developing nations. In November,

talks over the treaty broke down over the issue of how to measure nations' progress in reducing emissions. They are set to resume by midyear, after the Bush administration has formulated its position.

Doubters remain

Some argue that climate is too chaotic and complex to trust to any computerized prediction, or that Earth's climate is too stable to be greatly upset by a little more CO2. "I don't see how the IPCC can say it's going to warm for sure," says Craig Idso, a climatologist and vice president of the Center for the Study of Carbon Dioxide and Global Change in Tempe, Ariz. He calls predictions of drastic warming "a sheer guess" and says that extra carbon dioxide "is going to be nothing but a boon for the biosphere. Plants will grow like gangbusters."

But these skeptics appear to be losing ground. "There are fewer and fewer of them every year," says William Kellogg, former president of the American Meteorological Society and a retired senior scientist at the National Center for Atmospheric Research. "There are very few people in the serious meteorological community who doubt that the warming is taking place."

If the majority view holds up and temperatures keep rising, over the next century global weather patterns will shift enough to affect everyday life on every continent. The effects would vary wildly from one place to the next; what might be good news for one region (warmer winters in Fairbanks, Alaska) would be bad news for another (more avalanches in the Alps). Weather would become more unpredictable and violent, with thunderstorms sparking increased tornadoes and lightning, a major cause of fires. The effects of El Nino, the atmospheric oscillation that causes flooding and mudslides in California and the tropics, would become more severe. Natural disasters already cost plenty; in the 1990s the tab was $608 billion, more than the four previous decades combined, according to Worldwatch Institute. The IPCC will release its tally of anticipated effects on climate and societies on February 19 in Geneva. Key climate scientists say that major points include:

Death and pestilence. Cities in the Northern Hemisphere would very likely become hotter, prompting more deaths from heatstroke in cities such as Chicago and Shanghai. Deaths would also increase from natural disasters, and warmer weather would affect transmission of insect-borne diseases such as malaria and West Nile virus, which made a surprise arrival in the United States in 1999. "We don't know exactly how West Nile was introduced to the U.S., but we do know that drought, warm winter, and heat waves are the conditions that help amplify it," says Paul Epstein, a researcher at Harvard's School of Public Health (box, Page 50).

Wildfires. Rising temperatures and declining rainfall would dry out vegetation, making wildfires like last summer's—which burned nearly 7 million acres in the West and cost $1.65 billion—more common, especially in California, New Mexico, and Florida.

Rain and flooding. Rain would become more frequent and intense in the Northern Hemisphere. Snow would melt faster and earlier in the Rockies and the Himalayas, exacerbating spring flooding and leaving summers drier. "This is the opposite of what we want," says Gleick. "We want to be able to save that water for dry periods."

Rising sea levels. Sea level worldwide has risen 9 inches in the last century, and 46 million people live at risk of flooding due to storm surges. That figure would double if oceans rise 20 inches. The IPCC predicts that seas will rise anywhere from 3.5 inches to 34.6 inches by 2010, largely because of "thermal expansion" (warmer water takes up more space), but also because of melting glaciers and ice caps. A 3-foot rise, at the top range of the forecast, would swamp parts of major cities and islands, including the Marshall Islands in the South Pacific and the Florida Keys.

Water wars. Drought—and an accompanying lack of water—would be the most obvious consequence of warmer temperatures. By 2015, 3 billion people will be living in areas without enough water. The already water-starved Middle East could become the center of conflicts, even war, over water access. Turkey has already diverted water from the Tigris and Euphrates rivers with dams and irrigation systems, leaving downstream countries like Iraq and Syria complaining about low river levels. By 2050, such downstream nations could be left without enough water for drinking and irrigation.

Refugees. The United States is the single largest generator of greenhouse gases, contributing one quarter of the global total. But it, and other higher-latitude countries, would be affected less by climate change than would more tropical nations. The developing world will be hit hardest—and least able to cope. "Bangladesh has no prayer," says Stephen Schneider, a climatologist at Stanford University, noting that flooding there, and in Southeast Asia and China, could dislocate millions of people. "The rich will get richer, and the poor will get poorer. That's not a stable situation for the world."

Those daunted by this roster of afflictions will be cheered, a bit, by the United Nations group's report on how to fend off these perils, which will be released March 5 in Ghana. Not only is humanity not helpless in the face of global warming, but we may not even have to give up all the trappings of a First World lifestyle in order to survive—and prosper.

The first question is whether it's possible to slow, or even halt, the rise in greenhouse gases in the atmosphere. Scientists and energy policy experts say yes, unequivocally. Much of the needed technology either has already been developed or is in the works. The first step is so simple it's known to every third grader: Conserve energy. Over the past few decades, innovations from higher gas mileage to more efficient refrigerators to compact fluorescent lights have saved billions of kilowatts of energy. The second step is to use less oil and coal, which produce greenhouse gases, and rely more on cleaner energy sources such as natural gas and wind, and later on, solar and hydrogen. In Denmark, 13 percent of electricity now comes from wind power, probably the most economical alternative source. In Britain, a company called Wavegen recently activated the first commercial ocean- wave-energy generator, making enough electricity to power about 400 homes.

Taxing ideas

But despite such promising experiments, fossil fuels remain far cheaper than the alternatives. To reduce this cost advantage, most Western European countries, including Sweden, Norway, the Netherlands, Austria, and Italy, have levied taxes on carbon emissions or fossil fuels. The taxes also are intended to nudge utilities toward technologies, like coal gasification, that burn fossil fuels more cleanly. In Germany, where "eco-taxes" are being phased in on most fossil fuels, a new carbon levy will add almost 11 cents to the price of a gallon of gasoline.

But the United States has always shunned a carbon tax. John Holdren, a professor of environmental policy at Harvard's Kennedy School of Government, says such a tax could stimulate economic growth and help position the United States as a leader in energy technology. "The energy technology sector is worth $300 billion a year, and it'll be $500 to $600 billion by 2010," Holdren says. "The companies and countries that get the biggest chunk of that will be the ones that deliver efficient, clean, inexpensive energy."

A growing number of companies have already figured that out. One of the most advanced large corporations is chemical giant DuPont, which first acknowledged the problem of climate change in 1991. Throughout the past decade, the company worked to cut its carbon dioxide emissions 45 percent from 1990 levels. Last year, it pledged to find at least 10 percent of its energy from renewable sources.

Even more surprising was the dramatic announcement by oil giant BP in 1997 agreeing that climate change was indeed occurring. Even with other oil firms protesting that the evidence was too thin, BP pledged to reduce its greenhouse-gas emissions by 10 percent from 1990 levels by 2010. At the same time, BP Amoco is pouring money into natural gas exploration and investing in renewable energy like solar power and hydrogen.

Even America's largest coal-burning utility company is experimenting. American Electric Power of Columbus, Ohio, is testing "carbon capture," which would separate out carbon dioxide emissions and dispose of them in deep underground saline aquifers, effectively creating carbon-emission-free coal power. Application is at least a decade away. "If we're able to find creative solutions, they're going to place us at a competitive advantage in our industry," says Dale Heydlauff, AEP's senior vice president for environmental affairs.

In automobile manufacturing, there is already a race on for alternatives to fossil fuels. Several automakers like Ford, DaimlerChrysler, and Volkswagen have developed prototypes of cars run by hydrogen fuel cells rather than gasoline. The performance is very similar to that of today's cars, but the cost remains, for now, prohibitive. Fuel-cell vehicles are unlikely to be mass-produced until after 2010, and even then, people will need a push to make the switch. "Climate change is too diffuse to focus people's attention," says C. E. Thomas, a vice president at Directed Technologies, an Arlington, Va.,

engineering firm working on fuel cells. "But if we have another war in the Middle East or gasoline lines, that will get their attention."

Even with these efforts, and many more, climatologists point out that turning the atmosphere around is much harder than turning a supertanker. Indeed, atmospheric changes already underway may take hundreds of years to change. As a result, some vulnerable countries are already taking preventive, if costly, measures. More than half of the Netherlands lies below sea level and would be threatened by increased storm surges. Last December, the Dutch government outlined an ambitious plan to bolster the sea defenses. Over the next decade, the Netherlands will spend more than $1 billion to build new dikes, bolster the natural sand dunes, and widen and deepen rivers enough to protect the country against a 3-foot rise in ocean levels.

Some of the most successful adaptations to climate change probably won't involve high-tech gizmos or global taxes. They'll be as simple as the strips of cloth distributed to women in Bangladesh, which they use to screen cholera-causing microbes from water. Villages where women strained water have reduced cholera cases by 50 percent.

"Society is more robust than we give it credit for," says Michael Glantz, a political scientist at the National Center for Atmospheric Research. Like farmers who gradually change to new crops as wells grow dry, people may learn to live comfortably in a new, warmer world. ❖

Questions

1. The newest global-warming forecast is backed by data from what?
2. In the 1990s, how much was the tab for natural disasters?
3. What have most Western European countries done to reduce the cost advantage that fossil fuels have over the alternatives?

Answers are at the back of the book.

22

At a time when environmental concerns are at a high worldwide, the United States does not appear to have a consistent climate policy. Mounting public and international concern over the lack of a clear climate policy in the U.S. has prompted a call for the reframing of the debate over climate policy. To combat a stalemate in consensus over climate policies, the president should consider "no regrets" approaches to reducing greenhouse gases coupled with an improved understanding of climate change.

Climate Policy Needs a New Approach

David Applegate

Geotimes, May 2001

ENVIRONMENTAL GROUPS AND EUROPEAN LEADERS raised a hue and cry in March when President Bush reversed a campaign pledge to regulate power-plant emissions of carbon dioxide. At a March 29 press conference, Bush stated that "circumstances have changed since the campaign. We're now in an energy crisis." Bush linked his decision with the need to bring more natural gas resources to market before the power industry shifts to generating more electricity with natural gas rather than with carbon-intensive coal.

While pledging to work "with our allies to reduce greenhouse gases," Bush cautioned that he would not take any actions "that will harm our economy and hurt American workers." That statement reflects the new president's concern with the Kyoto Accords, signed by his predecessor in 1997 but never brought before the Senate for ratification. Sticking to a more oft-stated campaign promise, Bush has asked the State Department to determine how the United States would withdraw from the treaty, which set limits on carbon dioxide emissions for developed nations.

Repudiating the policies of one's predecessor is a time-honored approach, but it is not politically sufficient for an issue that the public perceives as a persistent problem. And

climate change is just such an issue. Public perception demands that Bush now articulate his own strategy for addressing climate change.

No Regrets

In mapping a new policy, the president would be well advised to reframe the issue in order to move beyond the current impasse in diplomatic and congressional negotiations. That means moving beyond carbon dioxide, which has proven to be a political non-starter.

Both sides of the climate debate accept that carbon dioxide levels are increasing, and there is fairly broad agreement that humans have been responsible for those increases in the past century. But beyond that, many politicians doubt any claims of a scientific consensus. Far more than President Clinton, President Bush will hear conflicting views on the link between carbon-dioxide levels and climatic warming—indeed, on whether warming is even happening and whether warming would be beneficial or harmful.

It makes sense to consider policy alternatives that produce benefits regardless of whether humans are indeed warming the planet. Such "no regrets" strategies help achieve policy objectives not directly related to climate change. The greatest advantage of "no regrets" policies is that they can be enacted without awaiting scientific agreement on human contributions to climate change or the magnitude and extent of future impacts of a changing climate. They are worth doing with today's climate. Policy-makers have proposed a number of such strategies, many focusing on reducing emissions of greenhouse gases and soot particles, which not only contribute to air pollution but also have a warming effect on the atmosphere.

What policy could be more appropriate in "no regrets" terms than one seeking to reduce current vulnerability to weather, both in the United States and around the world? Such vulnerability is increasing for reasons unrelated to climate change, but policies to improve resilience to today's weather would also make it easier for us to adapt to a future, warmer and wetter climate.

Resilient Today, Adaptable Tomorrow

Daniel Sarewitz—a former Geological Society of America Congressional Science Fellow—and Roger Pielke Jr. supplied one of the most convincing arguments for this approach in "Breaking the Global-Warming Gridlock," which appeared last July in The Atlantic Monthly. Sarewitz and Pielke do an excellent job of analyzing why the climate change debate has stalled over carbon dioxide, and then they propose how to reframe the debate in terms of reducing society's vulnerability to weather. As Sarewitz and Pielke point out, such an approach sidesteps the "determined and powerful opposition" that faces any efforts to reduce carbon-dioxide emissions.

Geologists often express frustration when politicians attribute individual weather events to global climate change. Along similar lines, Sarewitz and Pielke argue that disasters like Hurricane Mitch, which killed more than 10,000 people in Central America in 1998, "will become more common and more deadly regardless of global warming. Underlying the havoc in Central America were poverty, poor land-use practices, a degraded local environment, and inadequate emergency preparedness—conditions that will not be alleviated by reducing greenhouse-gas emissions."

In the United States, more and more people are moving into harm's way, bringing their wealth with them to coastal areas where they are vulnerable to hurricanes, floods and tsunamis. It is estimated that by 2025, three-quarters of Americans will live within 80 miles of a major coastline, up from 55 percent today.

The intensity of severe weather does not have to increase in order to cause greater and greater losses of life and prop-

erty. As William Hooke of the American Meteorological Society has noted, extreme events are the way that our planet conducts its business. The big difference in terms of impact is not the size of natural phenomena, but who is there to experience them.

While it is true that natural climate variations have been much greater in the distant past, a more relevant observation is that modern populations and attendant infrastructure have built up during a time of relative climatic stability. Whether human-induced or natural, climate change will test our adaptability and resilience. Sea-level rise, for example, could render coastal cities even more susceptible to damage from storms and flooding.

Sarewitz and Pielke note recent research showing that the West Antarctic Ice Sheet has been melting for thousands of years, thus limiting the role of human-induced warming as the "culprit." Such a finding does not eliminate the problem, but it does suggest the need for a different solution. Again, building resilience today will provide a long-term benefit, whether or not human-induced warming is to blame.

Sarewitz and Pielke recognize that the very mention of adaptation sends a chill down both sides of the climate change debate. To those who favor action to stop global warming, policies aimed at adaptation are tantamount to surrender. For those who see no threat, adaptation is unnecessary.

But stalemate is not inevitable, because "there is a huge potential constituency for efforts focused on adaptation: everyone who is in any way subject to the effects of weather." Sarewitz and Pielke go on to say that such a constituency could form the basis of a revitalized international climate change policy focused "on coordinating disaster relief, debt relief, and development assistance, and on generating and providing information on climate that participating countries could use in order to reduce their vulnerability."

Their article concludes: "As an organizing principle for political action, vulnerability to weather and climate offers everything that global warming does not: a clear, uncontroversial story rooted in concrete human experience, observable in the present, and definable in terms of unambiguous and widely shared human values, such as the fundamental rights to a secure shelter, a safe community, and a sustainable environment."

A Modest Proposal

The president should consider a strategy that couples policies for making society more resilient to natural hazards with policies for addressing climate change: to build better observational networks, improve our understanding of climate controls and impacts, and take "no regrets" approaches to reducing greenhouse gases. Many links already exist. Moreover, much of what we know about recent climate variability is because, 100 years ago, we set up hazard-observing networks and began to archive the data. The natural hazard mission provides a continuing stream of benefits and, simultaneously, builds a needed climate record.

A presidential policy initiative that includes all of these components promises progress at a time when current approaches to addressing climate change are at a standstill. And by addressing vulnerability, this policy could also bring us closer to that most universal of goals: help those who need it most. ❖

This article is reprinted with permission from *Geotimes*, Copyright American Geological Institute, 2001

Questions

1. By 2025, what percent of Americans will live within 80 miles of a major coastline?
2. What social factors contributed to the deaths of over 10,000 people by Hurricane Mitch in Central America in 1998?
3. Has the Kyoto Accord to reduce greenhouse gases ever been ratified by the U.S. Senate?

Answers are at the back of the book.

Massive subsidies have kept operation of landfills relatively cheap compared to the environmental costs of landfills. These subsidies have effectively prevented the composting industry from gaining a foothold in the waste management industry. While regulations currently allow landfills to defer environmental costs to future generations, the composting industry is forced to address present environmental concerns, which could stunt the growth of the composting industry. A new EPA proposal to deregulate most national standards for municipal landfills could have even more disastrous consequences for the practice of composting.

Bioreactors and EPA Proposal to Deregulate Landfills

Bill Sheehan, Jim McNelly

BioCycle, January 2003

DESPITE ITS SIGNIFICANT CONTRIBUTIONS as an environmentally preferable resource management technology, the composting industry has yet to show it can compete as an alternative to landfills. But this is only because current landfill regulations put composting at a severe disadvantage by allowing landfills to defer costs of inevitable environmental contamination to future generations while composting has to address environmental concerns in the short term before compost can be used beneficially.

A new proposal from EPA to deregulate most national standards for municipal landfills would make the present unacceptable situation even worse. The proposed rule would perpetuate landfilling and drive a stake in the heart of composting as a resource conservation practice, except in limited markets. Many in the environmental community have been lulled into thinking that landfills were safe and also were being increasingly deemphasized as a disposal option. Without a wake-up call, Americans may find their children and grandchildren opening their eyes to environmental clean-up costs far greater than the current Superfund program.

Failure of "Dry Tombs"

The ill-fated journey of the Mobro Garbage Barge in 1987 was supposed to have marked a watershed in which the future held great hopes for recycling and composting to finally be able to compete with landfilling. Impending "Subtitle D" rules by the Environmental Protection Agency (EPA) at that time were thought to herald a new era in which landfills would be highly engineered to prevent pollution, making them more expensive relative to composting.

Unfortunately, it didn't turn out that way. In a deliberate political decision to keep tipping fees low, the final landfill rules were fatally flawed. Consequently, the increase in composting has been small relative to the increase in waste production (especially in the last half of the 1990s), and it has had no substantial effect in reducing the amount of compostable organics still entering landfills (comprising more than 60 percent of landfill tonnage).

This outcome, however, did not result from a lack of practical alternatives. EPA could have required keeping biodegradable materials out of landfills, as is being done in Europe, parts of Canada, in San Francisco and several other cities in the U.S., where it is recognized that we are simply unable to safely manage decomposable material in the ground. Promoting source separation of food scraps, unrecyclable paper and yard trimmings for composting—just as the public was already successfully separating bottles, cans and newspaper for recycling—would have dramatically reduced multiple problems, including discharge of toxic leachate and generation of climate-changing and hazardous gases by landfills.

The reason the original flawed rule went into effect is because the practice of deferring liability for these discharges into the future made it appear that landfilling was less expensive than it actually is. The rule conveniently terminated liability for groundwater contamination 30 years after the landfill is closed, just when the elaborate system of barriers are expected to fail, thus saddling our grandchildren with the prospect of dealing with contaminated drinking water supplies at many locations. Had the EPA required incoming wastes to be treated and stabilized before burial, landfill tip fees would be $65/ton, instead of the $20/ton at many of today's megafills. This amounts to a subsidy of $45/ton.

"Even the best liner and leachate collection systems will ultimately fail," EPA's own technical staff noted in the Federal Register in 1988. Furthermore, insulating trash companies from competition and liability places the ultimate cost of clean-up on the public in what may become a "Super-Superfund" program costing hundreds of billions of dollars. EPA's own Inspector General wrote in the 2001 report, RCRA Financial Assurance for Closure and Post-Closure: "EPA officials acknowledge the lack of criteria or scientific basis for establishing the 30-year post-closure time frame. ... EPA made the decision to establish the time frame at 30 years, seemingly based on a compromise of these competing interests." (emphasis added)

"Bioreactors" a Risky Techno-fix

During the 1990s, EPA simply ignored landfill critics and composting interests. But gradually the facts became inescapable and it is now widely accepted even within the landfill and regulatory communities that the "dry tomb" approach to landfilling was fatally flawed. Instead of recognizing the futility of trying to manage organics mixed with toxics in huge mountains of garbage, however, EPA is promoting an extremely risky technological fix dubbed "bioreactors," that is, a technique of deliberately flooding the landfill with massive additions of liquids in an attempt to accelerate, rather than halt, decomposition.

Especially when done "on the cheap," however, efforts to accelerate decomposition will be only partially successful at resolving the longterm problems, and will create a whole new set of major short-term problems as well. For example, under proposed bioreactor designs, moisture levels would be increased from 20 percent to a range of 45 percent to 70 percent. By liquefying the waste load, this practice creates enormous engineering challenges that do not currently exist. Landfills would need to be engineered as reservoirs, not just dry fill. This has the effect of making soluble the complement of hazardous constituents that are a part of the trash. Accelerated decomposition also results in a rapid differential settlement of a heterogeneous landfill mass, thus destabilizing the support of the cover, and further increasing ingress of precipitation.

Moreover, the landfill structures that are supposed to contain this toxic slurry are not dug into the ground but are usually manmade mountains as much as 300 or more feet high. This mass and weight is contained by fragile sidewalk that are little more than a two-foot wide clay berm and plastic tarp. As one might expect, containing an unstable mass of tens of millions of tons of liquefied garbage high in the air behind such a frail barrier creates monumental management problems. There have already been several catastrophic landslides at test bioreactor sites attempting to reclaim and recirculate liquids.

This risky approach to landfill design is little more than a desperate attempt to maintain low landfill prices. The organic fraction of our waste stream ought to be treated as a valuable resource and composted, rather than mixed with toxics in garbage and buried, creating an insoluble problem. Tens of millions of dollars have already been committed by the EPA to advance the concept of landfill bioreactors. Not one penny, by contrast, has been invested in research on composting where methane generation can be controlled and a valuable product produced to boot.

Deregulating Landfills

EPA is doing more than investing in the risky techno-fix of bioreactor landfills. Even though the existing rules are too weak to protect the environment, EPA is now proposing a rule that would allow each state the sole discretion of issuing exemptions from most of those weak Federal Subtitle D regulations for landfill operators without any oversight by EPA of the granting of the variance.

EPA claims that the purpose of the proposed rule ("State Research, Development and Demonstration Permits for MSW Landfills") is to encourage innovation. However, in view of the fact that EPA already has federally supervised experimentation procedures in place and functioning, this claim can be seen for what it really is: a thinly disguised ruse to deregulate most of the minimum national standards for landfill permitting and open the flood gates to the most ill-considered bioreactor designs built on the cheap to insure that composting will not be able to compete in the 21st century. Instead of taking the common sense approach of preventing the problem by banning biodegradable material from landfills and investing in composting, EPA is bowing to the landfill industry which stands to profit from continuing to bury organics, because tight landfill supplies are the key strategy to acquire market power.

EPA's proposed rule represents a direct challenge to composting's economic future. Composters need to seize upon this attack as an opportunity to build support from the public and environmental groups. Our position is not only that the proposed, poorly conceived bioreactor loophole program be abandoned, but that EPA get serious about waste reduction and climate change by banning biodegradable material from landfills. Everyone of us is terribly busy, but few turning points are as critical for the long-term prospects of composters as this one. Please, before settling in to figuring how to make payroll this Friday, go to http: //www.GRRN.org/landfill/epa background.html for directions on how to submit your comments to EPA. ❖

Reprinted with permission from *BioCycle*, January 2003; www.biocycle.net

Questions

1. How long is a landfill liable for groundwater contamination after it is closed, and why is this a problem?
2. How much would landfill tip fees be if the EPA required incoming wastes to be treated and stabilized before burial? What are landfill tip fees at many of today's megafills?

3. What is "bioreactors"?

Answers are at the back of the book.

The international arms race may be over, but the race to clean up the environment after massive nuclear weapons production is still in full gear. After over a decade of planning efforts, the Department of Energy is finding the goal of environmental restoration to be inaccessible. Crowley and Ahearne suggest that the DOE would do better to recharacterize their definition of "clean up" to mean assessment and supervision of current and future environmental and health risks posed by the byproducts of the arms race.

Managing the Environmental Legacy of U.S. Nuclear-Weapons Production

Kevin D. Crowley, John F. Ahearne

American Scientist, November/December 2002

THE DEVELOPMENT OF THE ATOMIC BOMB during the Second World War was a stunning scientific and engineering achievement. Created by order of President Franklin Roosevelt in 1942, the Manhattan Project had produced by 1945 enough enriched uranium and plutonium for the August attacks on Hiroshima and Nagasaki that ended the Second World War. The sources of that enriched uranium and plutonium were the Oak Ridge site in Tennessee (then called Clinton Laboratories) and the Hanford Engineering Works in Washington, respectively—massive industrial complexes built from scratch to spearhead the development effort. After the war, these sites were expanded and additional sites established to meet the nation's growing demands for nuclear materials for the Cold War arms race with the Soviet Union.

The effort to build the bomb and win the arms race exacted a steep environmental price. Substantial quantities of radioactive and chemical contaminants were released to the environment, and even today large quantities of radioactive and toxic waste remain in storage at several sites. While it was being produced, few provisions were made for "solving" the problem of waste—that is, putting the waste into a stable

form and placing it out of reach of people and the environment—and the passage of time has since exacerbated the difficulties. With institutional memories fading and facilities aging, management of this legacy has become a more difficult, expensive and perilous undertaking.

The federal government, under the auspices of the U.S. Department of Energy (DOE), is making large expenditures of taxpayer funds to address the environmental legacy of the arms race. But after almost 13 years of effort and outlays of over $70 billion, the goal of "cleanup" is proving elusive. Through the work of independent expert committees appointed by the National Research Council, we have been following the DOE cleanup program for many years. Here, in a personal and unofficial assessment of what has been learned, we examine that program's efforts to come to terms with the environmental consequences of weapons production, including its technical and societal dimensions. Our examination suggests that it is time for the cleanup program to redefine success based on reducing and managing the human and environmental health risks that will extend far into the future.

Nuclear-Weapons Production

The U.S. nuclear weapons complex is massive in scale and highly dispersed: some 5,000 facilities located at 16 major sites and more than 100 smaller sites, ranging from mills to recover uranium from mined ore to facilities for weapons assembly, maintenance and testing. The largest facilities in the complex are those built for materials production and processing—Hanford, the Idaho National Engineering and Environmental Laboratory (INEEL), Oak Ridge and Savannah River in South Carolina. At these locations, nuclear materials (enriched uranium, plutonium and tritium) were produced for use in weapons, naval fuel and civilian nuclear applications. Their massive scale was dictated by the need for secrecy and safety as well as the physics of nuclear-materials production.

Manhattan Project scientists believed that an atomic weapon could be constructed using either of two radioactive isotopes, uranium-235 or plutonium-239. But obtaining sufficient quantities of these materials (tens of kilograms per bomb) proved a technically difficult and expensive challenge. Uranium-235, the principal isotope of a uranium weapon such as that used at Hiroshima, makes up about 0.7 percent by mass of natural uranium, most of the remainder being uranium-238. To be usable in a fission weapon, uranium-235 must be concentrated (enriched) to greater than 20 percent and preferably, for greatest efficiency, to over 90 percent. Uranium enriched to 20 percent or more uranium-235 is "highly enriched uranium" (HEU).

Uranium enrichment during the Manhattan Project exploited the small mass differences between uranium-235 and uranium-238 using electromagnetic separation and gaseous diffusion—the latter a process still used in the United States to enrich uranium for civilian power plants. Thousands of separation stages were required to obtain sufficient uranium-235 enrichment, and the facilities housing these processes were scaled to yield the required quantities.

Two plants were built during the Manhattan Project to obtain the roughly 50 kilograms of enriched uranium for the "Little Boy" weapon that was dropped on Hiroshima: the K-25 building at the Oak Ridge Gaseous Diffusion Plant and the Y-12 electromagnetic-separation plant at Oak Ridge. After the war, the Atomic Energy Commission (AEC) expanded the gaseous diffusion plant and built two additional enrichment facilities, the Paducah plant in Kentucky and the Portsmouth plant in Ohio, to meet growing civilian demands.

Plutonium-239, the principal isotope in "weapons-grade" plutonium, was produced by irradiating uranium-238 fuel rods with neutrons in large production reactors built along the Columbia River at the Hanford site. The capture of a neutron by uranium-238 produces uranium-239, which subsequently decays to neptunium-239 and then plutonium-239.

Once a plutonium-239 atom is produced, additional neutron captures can cause it to fission or produce heavier "reactor grade" plutonium (primarily plutonium-240 and -241), which is not as suitable for weapons. To minimize additional captures, the Hanford operators removed the fuel rods from the reactor after a short irradiation time, but this required large throughputs of uranium to obtain the needed quantities of plutonium. Three reactors and two chemical processing plants were built at Hanford to produce plutonium for the war effort. Following the war, the AEC built six additional reactors and three chemical processing plants at Hanford and five reactors and two chemical processing plants at Savannah River to meet the growing plutonium demands.

Tritium, an isotope of hydrogen and a key component of fusion weapons ("hydrogen bombs"), which were first developed by the United States in the early 1950s, was produced by the irradiation of lithium-6 targets in the production reactors at Savannah River. An isotope-separation facility was constructed at the Y-12 site at Oak Ridge to produce lithium-6 for this purpose.

During the roughly 45 years of nuclear-materials production in this country, about 103 metric tons of weapons-grade plutonium were obtained from the production reactors at Hanford and Savannah River; 994 metric tons of highly enriched uranium were obtained from the enrichment plants at Oak Ridge, Portsmouth and Paducah. Some of these materials have been declared to be surplus to U.S. defense needs (see table). Plans are now in place to turn most of this excess into fuel for use in nuclear power plants. The spent fuel will be disposed of in a geologic repository. Dealing with the by-products of nuclear-materials production is another matter.

Environmental Consequences

Although the production of nuclear materials generated huge quantities of waste, good records of radioactive and chemical waste production and environmental discharges generally were not kept until the 1970s. What is known of that early history today is based on reviews of written records supplemented by process knowledge and mass-balance calculations. We have been selective in our use of data in the following discussion, preferring more recent sources that contain documentation for the estimates. We also have rounded the data except where there is demonstrated support for greater precision.

The uranium-enrichment process produced low-level solid and liquid wastes and other process liquids, and up to

200 kilograms of depleted uranium (enriched in uranium-238) for every kilogram of HEU. There are on the order of half a million metric tons (metal equivalent) of depleted uranium in storage at several sites, and although this material is not classified as waste, most of it has no agreed-upon disposition pathway.

The separation of lithium-6 for tritium production used on the order of 10,000 metric tons of mercury, of which about 900 metric tons is unaccounted for. The DOE estimates that about 110 metric tons was discharged into East Fork Poplar Creek at Oak Ridge, and some of this contamination has migrated offsite and into the Clinch River–Watts Bar Reservoir system that is used for recreation and municipal water supply. The inorganic mercury compounds in this waste are not thought to be toxic, but they can pose a hazard to human beings if transformed to methylmercury by soil and water microorganisms. Plutonium production also produced large quantities of waste: During the 50 years of operation of the Hanford site, for example, about 67 metric tons of plutonium were produced from almost 97,000 metric tons of irradiated uranium fuel. Chemical processing of that uranium to recover plutonium produced some 2 million cubic meters (500 million gallons) of highly radioactive, chemically toxic waste, and another 1.7 billion cubic meters (450 billion gallons) of process liquids. The DOE reports that about 76,000 cubic meters of solid waste contaminated with actinides (principally plutonium) and an additional 1.2 million cubic meters of other solid low-level wastes have been buried at the site.

Some of the waste generated by nuclear materials production was released directly to the environment. Volatile gases from chemical processing were vented directly into the atmosphere, sometimes without filtering, especially during the early years of production. Reactor cooling water contaminated with conditioning chemicals such as chromium and with radioactive isotopes produced in the reactor (neutron-activation products) were also discharged. Waste liquids were discharged into large surface ponds or into subsurface soil and groundwater through injection wells and other drainage structures. Radioactive and chemically contaminated solid waste was burned or dumped into shallow pits and trenches.

Indeed, there are thousands of "release sites" that are current or potential future sources of contaminant releases to the environment. As a result of such releases, soil and groundwater at many sites are extensively contaminated with industrial solvents, toxic chemicals, metals and radionuclides.

Large volumes of waste remain in storage at several sites and could become significant sources of future environmental contamination if not managed properly. At Hanford, for example, there are about 200,000 cubic meters of high-level waste in storage in 177 large underground tanks. These tanks have been in service between 16 and 58 years; under current plans, the last tank will not be closed until about 2046. The older, single-containment tanks were designed with service lives of 10 to 20 years, although no one really knew how long they would last. One tank began leaking just six years after it was put into service, and to date 67 of these tanks are suspected to have leaked up to 5,700 cubic meters, or 1.5 million gallons, and possibly more than a million curies of high-level waste into the subsurface. (Curies are a measure of radioactivity in a material; for comparison, a ton of uranium-238 has 0.3 curies.) Some of this contamination has reached groundwater.

At Savannah River, there exist some 130,000 cubic meters (34 million gallons) of high-level waste stored in 48 underground tanks. Nine of the tanks have leaked waste into their secondary containments, and a few tens of liters of waste leaked into the environment from one tank when the secondary containment overflowed. Efforts are now under way to immobilize the sludge fraction of this waste in a borosilicate glass matrix.

The Idaho site processed naval spent fuel and some research reactor fuel to recover enriched uranium, but here, unlike at Hanford and Savannah River, the high-level waste was immobilized as a powdered ceramic (calcine), about 4,000 cubic meters of which are being stored in stainless steel bin sets inside steel-reinforced concrete silos. These structures were designed to contain the waste for up to 500 years. Additionally, another 4,000 cubic meters of so-called "sodium-bearing waste" liquids await disposition in some of the site's 11 underground storage tanks.

Past practices for managing the large volumes of waste generated by nuclear materials production, when judged by today's standards, appear ill-informed at best, bordering on reckless at worst. It is important, however, to judge these practices against the prevailing environmental attitudes and practices during the Second World War and Cold War. The Manhattan Project was created during a national emergency at a time when the future of Europe and Asia hung in the military balance. National priority was given to weapons production at the expense of waste management. This sense of urgency, and a shroud of secrecy that hid production activities from public view, carried over into the Cold War, although increasing effort was given to minimizing environmental releases as time went on.

Wartime shortages of materials such as stainless steel created further difficulties. Carbon steel was employed to construct the waste tanks at Hanford and Savannah River, with the result that the high-level waste, which was highly acidic, had to be neutralized with alkaline chemicals such as sodium hydroxide to reduce tank corrosion. The addition of these chemicals to the waste increased volumes and produced solid precipitates. Later chemical processing and volume-reduction operations to reduce radioactive heat generation and conserve tank space further increased physical and chemical heterogeneity. Characterization of this waste and removal of the precipitates from the tanks will be difficult and expensive, especially at Hanford.

The waste-management decisions made during the Manhattan Project and ensuing Cold War created the environmental problems that the nation now confronts. These decisions continue to exact a steep price, both in the high annual costs of managing the stored waste and environmental contamination, and also in the loss of trust by citizens in their government as the consequences of waste-management practices carried out in secrecy for almost five decades have become public knowledge.

From Production to "Cleanup"

The decline of large-scale nuclear-weapons production began in the late 1970s and accelerated through the 1980s, coinciding with the thaw in Cold War relations that culminated in the Strategic Arms Reduction Treaty (START) and the breakup of the Soviet Union, both in 1991. At the same time, the reactor accidents at Three Mile Island in 1979 and Chernobyl in 1986 raised public concerns about the continuing operations of U.S. production reactors. In May 1986, Energy Secretary John Harrington asked the National Academy of Sciences and National Academy of Engineering to review the safety of the government's production and research reactors. He also commissioned a group of experts to review the operation of the N-Reactor at Hanford. Based on that review, he shut down the reactor in 1987, commenting that the United States had no need for it because the country was "awash in plutonium."

During this same period states also were beginning to assert their authority to regulate environmental releases at the sites, prompted by a 1984 federal court ruling that the Y-12 site at Oak Ridge was subject to state regulation under the Resource Conservation and Recovery Act. Complaints from Colorado led to the June 1989 Federal Bureau of Investigation raid and closure of the Rocky Flats site, a 1951-vintage weapons-component manufacturing facility near Denver, for violations of federal environmental laws. Five months later, Energy Secretary James Watkins announced the creation of the Office of Environmental Restoration and Waste Management (now the Office of Environmental Management)

and declared a new mission for weapons sites: environmental cleanup. The era of large-scale nuclear-weapons production had ended.

The new cleanup program contrasted, in many respects, with the production operations. From the earliest days of the Manhattan Project, weapons production had been conducted with scientific and technical rigor and a strong focus on meeting production goals that were noticeably lacking in the early years of the cleanup effort. One of the first actions taken by the new program, before it had developed an adequate understanding of the environmental insults at its sites or its scientific and technical capabilities to address them, was to negotiate legally enforceable cleanup agreements with states and regulators. Many of these original agreements had to be renegotiated after the problems were more fully understood. At present, the cleanup program is operating under some 70 separate agreements that contain more than 7,000 schedule milestones, many of which are potentially enforceable via court action.

Although the cleanup program has been in operation for over a decade, it has, until recently, accomplished relatively little actual cleanup. To be sure, DOE has had some important recent successes, both in site remediation and waste disposal. Perhaps its most notable waste disposal success was the 1999 opening of the Waste Isolation Pilot Plant near Carlsbad, New Mexico. This deep geologic repository will eventually be used to dispose of up to about 175,000 cubic meters of defense-generated transuranic waste (mainly plutonium-contaminated debris, clothing and tools, and the like) from nuclear weapons sites. Additionally, DOE has recommended Yucca Mountain, Nevada as the site for a deep geologic repository for spent fuel and high-level waste and is now in the process of developing an application for a construction license, which it plans to submit to the Nuclear Regulatory Commission in 2004. If constructed, this repository will be used to dispose of the immobilized high-level waste and spent fuel from nuclear weapons sites along with commercial spent fuel.

The notable remediation successes include the stabilization and capping of mill tailings piles and the cleanup of some Manhattan-era sites, the latter of which is presently being carried out by the Army Corps of Engineers. Also, successful efforts are being mounted at many sites to characterize the nature and extent of environmental contamination, halt the spread of contaminated groundwater, and cap waste burial sites to retard water infiltration and contaminant leakage. Work also is proceeding to decontaminate and demolish buildings and clean up contaminated soil and groundwater at some of the smaller sites (such as Fernald and Mound) so that they can be declared closed around 2006.

Perhaps the most significant technical success in the remediation program to date has been the construction and successful operation of a $2.5 billion plant at Savannah River for immobilizing high-level waste, which went into production in 1996 and has to date produced more than 1,200 canisters of borosilicate waste glass. At Hanford, work also has begun to cocoon the nine production reactors, remediate contaminated soil and groundwater along the Columbia River, and stabilize corroding spent fuel from the N-Reactor that has been stored for over decade in two unlined water basins next to the river, one of which is leaking. This fuel is being dried, canned and placed into temporary storage away from the river. With the notable exception of the immobilization program at Savannah River, however, none of these remediation actions has been technically demanding. In fact, attempts to undertake the technically demanding tasks have failed, due largely to inadequate scientific and technical understanding. Three examples serve to illustrate this point.

In the early 1990s, the Idaho laboratory began a project to excavate and treat waste and contaminated soil from a 1-acre site known as "Pit 9," one of a series of pits and trenches used for disposal of low-level and transuranic radioactive waste. Pit 9 is thought to contain about 7,000 cubic meters of sludge and other solids contaminated with plutonium from Rocky Flats, and the remediation effort was designed to demonstrate retrieval and processing technologies that could be applied elsewhere on the site. The DOE awarded a $200 million contract for this work in late 1994, but the project fell behind schedule, and costs exceeded the contract price before any waste had been retrieved or processed. The contract has been canceled, and the contractor has alleged that inadequate characterization of the waste in the pit contributed to this failure. Excavation of waste from this pit may not take place until 2004, fully a decade after the initial contract was awarded.

Efforts are now under way at Savannah River to develop a chemical process to remove radionuclides, principally cesium, strontium and plutonium, from the nonsludge fraction of its high-level waste for immobilization in glass. Savannah River contractors spent 10 years and almost $500 million to develop an in-tank precipitation process for removing cesium, but when this process was placed in production in one of the underground storage tanks, large quantities of benzene, an explosive hazard, were generated. Subsequent investigations and experiments failed to positively identify the benzene-generation mechanism, although a catalytic reaction involving trace elements in the waste was thought to be responsible.

DOE–funded scientists are now developing a solvent-extraction process that has a high selectivity for cesium. This process looks promising, but the schedule for waste retrieval and processing has been set back several years. The delay would likely have been much longer if not for the foresight of the department's research and development organizations, which funded research on alternative separation processes before the problems became evident.

There have been several attempts at Hanford, starting in the early 1990s, to begin retrieving and immobilizing high-level waste from its tanks using approaches similar to those at Savannah River. Construction of a facility to immobilize about 10 percent by volume and 25 percent by radioactivity of the liquid high-level waste finally began this year with the start of construction of a waste treatment and immobilization facility. This phase-1 project is slated to last until 2018 and cost about $15 billion. Hanford has not yet determined how it will process the remaining waste, or how it will retrieve the solid or semi-solid wastes from its single-containment tanks to meet the 99 percent removal milestone required by its compliance agreement with the State of Washington. Retrieval of this waste without damaging the tanks and releasing contaminants to the environment may be difficult using currently available technologies.

To be fair, the DOE has tried several times in recent years to improve the effectiveness of the cleanup effort. In 1995, the assistant secretary for environmental management announced a "10-year plan" for reducing the high annual carrying costs of the sites by accelerating the closure of smaller sites. Several sites, including the Mound (Ohio) and Rocky Flats sites, are now slated to be closed by 2006. The current administration is developing a plan to "accelerate cleanup" by focusing on risk reduction and negotiating with site regulators to shorten cleanup schedules. The objective is to reduce the current $220 billion to $300 billion estimated life-cycle cost of the cleanup program by $100 billion and 40 years.

Although these goals strike us as sensible, the success of this effort will hinge on several factors. Will regulators be willing to modify their compliance agreements with the DOE? Will state and local authorities and the site administrators themselves allow reallocation of budgets so that high-risk projects can be funded on an accelerated schedule—or conversely, will Congress allocate additional funds for this purpose? Will the DOE and contractors exercise good judgment in developing and applying remediation plans and, especially, learn from past experiences at the sites to avoid repeats of some of the problems described previously? Can DOE-funded investigators come up with timely solutions when new problems are identified, as they did for the cesium-separation problem at Savannah River?

Coming to Terms with "Cleanup"

The term "cleanup" poorly describes the current activities at DOE sites: Only a small portion of the approximately $7 billion in annual funding is actually used for contaminant removal and waste processing. Most of the budget is spent on site surveillance and maintenance. The cleanup program refers to these surveillance and maintenance costs as "mortgage costs."

In our view, these high mortgage costs are slowing work on high-risk problems that, if not addressed in a timely fashion, could lead to nasty future surprises. The slow progress in remediating the high-level tank wastes at Hanford and Savannah River is of particular concern. Many of the tanks are now well beyond their design lives and contain chemically complex and highly toxic waste, much of it in a liquid state. Some of the tanks are now leaking, and the number of "leakers" is likely to increase as the tanks age. Accidents, acts of God and terrorism are also concerns as long as the liquids remain in the tanks. Under current schedules, it will be several decades before all of this waste is recovered and immobilized, and some of the hardest work (such as retrieval of the "bottoms," rich in transuranic elements such as plutonium, from the single-containment tanks at Hanford) is being deferred until the later stages of the remediation effort.

The use of the term "cleanup" also suggests that the primary objective of the program is to remove waste and environmental contamination and return the sites to other productive uses. In fact, although some sites or parts of sites can be cleaned up and released for other uses, sometimes with few or no restrictions, the DOE has acknowledged that this will prove to be the exception rather than the rule and that parts of more than 100 sites are expected to be unacceptable for unrestricted release after cleanup. At many sites, and especially the large ones, contaminants are too widely dispersed in the environment to be recovered with current technologies. The stored wastes that exist at these sites can (and should) be processed to reduce volumes and stabilize the hazardous constituents, but after processing much of this waste will be reburied at the site. The hazards will be reduced or relocated, but not eliminated.

This fact is well recognized within the program, which defines cleanup as the completion of those actions necessary to meet agreed-upon standards and objectives, and not necessarily the removal of all waste and contamination. The expectations of regulators and local communities for achieving contaminant reduction and waste removal have been moderated since the cleanup agreements were signed as the technical difficulties and high costs of progress have become apparent. We sense, however, that expectations may still be higher than warranted in view of the difficult problems ahead, especially for the remediation of burial pits and trenches (such as Pit 9) and the retrieval and processing of high-level waste from the underground tanks, especially at the Hanford Site. Past success in site cleanup may not be a good harbinger of future prospects, because most of the difficult and costly problems have yet to be tackled.

Reducing and Managing Risk

Since its creation in 1989, the cleanup program has focused on developing and executing negotiated milestones to achieve specific cleanup tasks or levels of contaminant reduction, while at the same time (according to some critics) maintaining high levels of employment at sites that no longer have a national defense mission. It is becoming clear that many of these "activity-based" milestones may not be achievable with current technologies. Furthermore, the milestones are not designed around goals of protecting human and environmental health. If achieving such protection is the ultimate goal of the cleanup program, we believe that it may make more sense to organize major programmatic milestones around agreed-to levels of risk reduction without specifying in advance the specific remedial actions to be taken to achieve those reductions.

The judicious use of "risk-based" milestones could have several benefits. Such milestones could, for example, provide a better measure of progress and encourage the investment of funds where the greatest risk reductions could be achieved. They also could encourage greater creativity in the selection of "end states" for cleanup and the remedial actions to achieve them, creativity that is lacking in current activity-based milestone approaches.

Of course, the use of a risk-based approach requires that risk estimates be developed for site hazards. The cleanup program has had difficulty developing a risk-based analysis—the sites are complex, not all of the contaminants (groundwater plumes, for instance) have been located and characterized, nor is all of the waste adequately characterized. The cleanup program does not even use the "risk" concept consistently: Sometimes risk is defined based on effects on the health of off-site populations only, not including on-site workers, and other times risk is defined programmatically, that is, whether a particular action can be completed on time and within budget.

Long-Term Stewardship

As the title of this article suggests, a great deal of the environmental legacy of nuclear-weapons production may end up being managed, not eliminated. Many sites, or portions of sites, will not be remediated to levels deemed adequate for unrestricted access, and either the federal government or a state or local government will become landlords of last resort, with attendant responsibilities for protecting public and environmental health. In some cases this protection will come in the form of long-term surveillance to guard against human access or further environmental releases, and in other cases active measures such as groundwater treatment will be required. Some of these responsibilities may last indefinitely.

These long-term responsibilities have received little consideration by the cleanup program until recently, and even now "long-term stewardship" of contaminated sites is viewed as a separate activity from cleanup. Yet there is a very real trade-off between cleanup and stewardship—that is, protection against a hazard can be achieved either by eliminating it outright (through cleanup), managing it until it ceases to be hazardous (long-term stewardship) or a combination of both approaches. Over the short term, hazard management is usually less difficult and expensive than hazard elimination, but the long-term costs are not clear, and the effectiveness of long-term stewardship depends to a great extent on the continuing willingness and ability of future generations and institutions to manage the hazard, factors over which the current program has little or no control.

Given this trade-off between cleanup and stewardship, we suggest that both choices need to be put on the table when deciding on end states and remedial actions to achieve them, fully recognizing that a reliance on stewardship places a heavier burden on future generations. The use of risk-based cleanup approaches described earlier would help make these choices explicit.

There may be good reasons for relying on stewardship in some instances, especially if cleanup is not technically feasible or cost effective. Indeed, society routinely makes this choice for managing other kinds of waste, including chemically hazardous waste, although there is little or no evidence to demonstrate its effectiveness over multiple generations, and much evidence to the contrary.

Under current regulatory regimes, decisions to rely on long-term stewardship must be revisited periodically, and further actions to reduce hazards made if necessary and feasible. We believe that the ultimate success of long-term stewardship as a solution to the waste problem at DOE sites will hinge on advances in science, especially those elements of the social sciences that bear on the effective design and operation of durable institutions. There is reason to be hopeful given the rapid advances in the five decades since the Manhattan Project; yet continuing investments in building scientific and institutional capacities are essential to ensure the continued protection of people and the natural environment around these sites. The cleanup program is planned to last for several decades even under the most optimistic scenarios—consequently, wise research and development investments made today will likely pay great future dividends.

Acknowledgments

The authors are grateful for the assistance of Allen Croff (Oak Ridge National Laboratory), Kai Lee (Williams College) and Chris Whipple (ENVIRON International, Inc.), who provided information for this article and review comments on an earlier draft; Thomas Wood (Idaho Engineering and Environmental Laboratory), who provided photographs; Bruce Napier (Pacific Northwest National Laboratory), who provided information on environmental releases at Hanford; and Roy Gephart (PNNL), who provided information, comments and photographs.

Bibliography

Consortium for Risk Evaluation with Stakeholder Participation. 1999. *Peer Review of the Department of Energy's Use of Risk in its Prioritization Process*. New Brunswick, N.J.

Gephart, R. E. In press. *Hanford: A Conversation*. Columbus, Ohio: Battelle Press.

National Research Council. 1987. *Safety Issues at the Defense Production Reactors*. Washington: National Academy Press.

National Research Council. 1996. *Understanding Risk: Informing Decisions in a Democratic Society*. Washington: National Academy Press.

National Research Council. 2000. *Long-Term Institutional Management of U.S. Department of Energy Legacy Waste Sites*. Washington: National Academy Press.

National Research Council. 2000. *Research Needs in Subsurface Science*. Washington: National Academy Press.

National Research Council. 2000. *Alternatives for High-Level Waste Salt Processing at the Savannah River Site*. Washington: National Academy Press.

National Research Council. 2001. *Science and Technology for Environmental Cleanup at Hanford*. Washington: National Academy Press.

Russell, M. 1998. Toward a productive divorce: Separating DOE cleanups from transitional assistance. *Annual Review of Energy and Environment* 23:439–63.

U.S. Department of Energy. 1995. *Risks and the Risk Debate: Searching for Common Ground* (3 volumes). Washington: Office of Environmental Management.

U.S. Department of Energy. 1995. *Closing the Circle on the Splitting of the Atom*. Washington: Office of Environmental Management.

U.S. Department of Energy. 1996. *Plutonium: The First 50 Years*. DOE/DP-0137. Washington: Office of Defense Programs.

U.S. Department of Energy. 1997. *Linking Legacies: Connecting the Cold War Nuclear Weapons Production Processes to Their Environmental Consequences*. DOE/EM-0319. Washington: Office of Environmental Management.

U.S. Department of Energy. 1998. *Accelerating Cleanup: Paths to Closure*. DOE/EM-0362. Washington: Office of Environmental Management.

U.S. Department of Energy. 1998. *Commercial Nuclear Fuel from U.S. and Russian Surplus Defense Inventories: Materials, Policies, and Market Effects*. DOE/EIA-0619. Washington: Energy Information Administration.

U.S. Department of Energy. 2000. *A Strategic Approach to Integrating the Long-Term Management of Nuclear Materials: A Report to Congress*. Washington.

U.S. Department of Energy. 2001. *Summary Data on the Radioactive Waste, Spent Nuclear Fuel, and Contaminated Media Managed by the U.S. Department of Energy*. Washington: Office of Environmental Management. ❖

Reprinted with permission of *American Scientist*, Magazine of Sigma Xi, the Scientific Research Society.

Questions

1. What two reactor accidents raised public concern about the continuing operations of U.S. production reactors?

2. What is Yucca Mountain?

3. What is the $7 billion in annual funding used for at DOE sites?

Answers are at the back of the book.

25 Global warming has been affecting the behavior and migratory range of many North American bird species. The American Bird Conservancy forecasts that the initial changes that have already been observed in birds are an omen of increasing changes to come. Diversity in bird species has been on the decline, and the downward spiral is in danger of gathering momentum if the environment continues to be degraded. Without meaningful action many bird species will be forced either into extinction or out of their habitats; and the results of waning bird populations could be disastrous for the environment.

Silent Spring: A Sequel?

Les Line

National Wildlife, December 2002/January 2003

BY THE LATE 21ST CENTURY, says scientist Jeff Price, the distribution of North American birds will almost certainly look nothing like the species' range maps found in today's popular field guides. Some startling possibilities: State birds such as the Baltimore oriole (Maryland) and black-capped chickadee (Massachusetts) will have vanished from their official residences; there will be painted buntings in southern Minnesota but no bobolinks; the golden-cheeked warbler of Texas Hill Country will be extinct, along with other endangered songbirds whose habitat is sharply limited; and Neotropical migrants like the Cape May warbler will have moved farther north, leaving the southern boreal forest more vulnerable to devastating outbreaks of spruce budworms, which the birds now feast on during the breeding season.

Price, director of climate change impact studies for the American Bird Conservancy, is coauthor with NWF researcher Patricia Glick of a recent report, *The Birdwatcher's Guide to Global Warming*. The report's projections of avifaunal chaos are based on his computer models, which assume a doubling of atmospheric concentrations of carbon dioxide (CO_2) over levels found in preindustrial times, as well as substantial increases in other heat-trapping greenhouse gases. According

to the world's leading climatologists, such changes may happen sometime over the next 50 to 100 years. And although the past decade was the warmest on record so far, scientists also project average global temperatures will rise another 2.5 to 10.4 degrees F by the year 2100. "That's ten times faster than the sustained rate of temperature change since the last ice age," says Price.

Already, there are signs that birdlife is responding to climate change. In an analysis of three decades of data published in *Nature* last spring, researchers reported that a dozen British bird species have shifted their ranges an average of 12 miles north over the past 20 years. In North America, summer distributions of many Neotropical migrants also are shifting northward. The biggest change so far involves the golden-winged warbler, a strikingly patterned denizen of brushy clearings. The center of its range has moved nearly 100 miles north over the past two decades, says Price.

Warming also has had a measurable impact on the timing of such seasonal events as migration and breeding. According to the *Nature* study, numerous European and North American bird species are migrating an average of nine days earlier and breeding ten days sooner than they did 30 years ago. A long-term study on Michigan's Upper Peninsula found that 15 spring migrants, including the rose-breasted grosbeak and black-throated blue warbler, were arriving up to 21 days earlier in 1994 than they had in 1965. Price and his colleagues say such changes are merely hints of what to expect as temperatures continue to rise.

To map the potential summer distribution of North American songbirds in the future, Price developed large-scale statistical models of how climate variables such as average seasonal temperature and precipitation determine a species' present range. He then applied to these models Canadian Climate Center projections of conditions that may exist when atmospheric CO2 levels double. His results, summarized in the report, are both dramatic and troubling.

Price found that Great Lakes states, for example, may lose 53 percent of the Neotropical migrant songbirds that presently nest in the region. Other projected losses include 45 percent fewer Neotropical migrants in the Mid-Atlantic region, 44 percent fewer in the northern Great Plains and 32 percent fewer in the Pacific Northwest. Some losses could be offset by new species moving into an area. The Southwest may gain "some great new Mexican birds," Price notes. Yet overall avifaunal diversity is expected to decline.

Meanwhile, warming also may affect songbird habitat. Because native forests are adapted to local climates, many trees acclimated to cool environments are likely to shift northward. In New England, for example, southern oaks and hickories may replace today's mix of maple, birch and beech trees. Subalpine spruce-fir forests could die out as mountain habitats warm, and subalpine meadows in the Rocky Mountains will probably disappear. As the Southwest becomes moister, desert landscapes could be transformed into grasslands and shrublands.

The impact of these new distributions—of both birds and the ecosystems they inhabit—could be complicated by continued changes in the animals' migration and breeding behavior. For some species, climate change may actually turn out to be a good thing, at least in the short term.

Consider Virginia's prothonotary warbler. In swamp forests along the James River, Virginia Commonwealth University biologist Charlie Blem has been inventorying this species for nearly two decades. He reports that the birds, golden orange beacons in a gloomy forest, have been arriving from their Caribbean and South American wintering grounds an average of one day earlier each year since 1987. (Last spring the first male warblers were sighted on April 3 compared with April 21–23 in the project's early years.) During the same period, the average springtime temperature at Richmond's airport has increased nearly two degrees. "I

predict that if this trend keeps up, the species will become a year-round resident of North America," Blem says.

The warblers also are thriving. Blem has erected 316 nest boxes for the birds along a 19-mile swampland trail that takes his team of students and volunteers a week to cover by canoe—counting eggs, banding nestlings and capturing adults. Though the species has been declining in many of its Southeast strongholds, James River warblers are "nesting earlier, incubating more eggs, raising as many as three broods and surviving longer," says Blem.

The downside of early migration and reproduction, of course, is that a species' breeding cycle could get out of sync with its food supply. "As a result," says Price, "early birds may not get the worm." At the Rocky Mountain Biological Laboratory in Colorado, for example, robins have been arriving an average of two weeks earlier in spring than they did in the late 1970s, apparently in response to warmer temperatures at their low-altitude wintering grounds. But when robins reach the mountaintop now, it is still winter, and the birds must wait longer for snow to melt before they can feed. This puts the birds and their offspring at greater risk of starvation.

Other ecosystem connections are threatened by warming as well. According to biologist Terry Root of the Stanford University Center for Environmental Science and Policy, "well-balanced bird communities as we know them will likely be torn apart. As species move, they may have to deal with different prey, predators and competitors as well as habitats that are less than ideal."

Many birds, Root reminds us, are linked to specific vegetation—and it could take decades or centuries for plants to respond to global warming. "The endangered red-cockaded woodpecker of the Southeast needs mature pine forest, habitat that is already scarce. If the species has to shift its range northward to stay in a cooler environment, can it adapt to use other trees?" she asks.

If an ecosystem, in turn, loses a bird that helps control insect pests, the results could be catastrophic, for humans as well as other species. In the boreal forests of eastern North America, for instance, nesting wood warblers are important predators of the eastern spruce budworm, which defoliates millions of acres of timberland every year. Without the birds, those losses would likely be far greater. Under normal conditions, warblers consume up to 84 percent of the budworm's larvae and pupae.

Price's models, however, project that with a doubling of CO_2, three of the most significant budworm predators—Tennessee, Cape May and blackburnian warblers—may disappear from spruce-fir forests south of Hudson Bay. This "decoupling of predator and prey," he warns, could lead to large-scale pesticide use if budworm outbreaks get severe.

Similar problems could occur in the West, where savannah sparrows, sage thrashers and other species that help control rangeland grasshopper populations are expected to move north. "A single pair of savannah sparrows raising their young consumes an estimated 149,000 grasshoppers over the breeding season," says Price. "Unless all of the components of this ecosystem—grasslands, insects and birds—change at the same time, an unlikely prospect, we're looking at more grasshopper outbreaks in the future."

For birds, the most devastating consequence of global warming would be loss of entire habitats on which species depend—and, unfortunately, scientists say such losses are a very real possibility. On the East Coast, for example, a rising Atlantic Ocean would inundate coastal marshes and beaches critical to the survival of 150 species of shorebirds, waterfowl and other birds. Each spring, the shores of Delaware Bay alone provide a critical food bonanza for about a million migratory shorebirds that feast on the eggs of horseshoe crabs that spawn there. Drops in the bay's horseshoe crab population, caused in part by habitat loss, have already resulted in declines of up to 50 percent among some of these species, including sanderlings, red knots and ruddy turnstones.

Droughts across the continent's northern prairies, meanwhile—also projected by climate models—could dry up tens of thousands of "prairie potholes," vast seasonal wetlands that are crucial to millions of waterfowl throughout the breeding season. Loss of these wetlands would decimate populations of mallards, blue-winged teal, gadwall and other ducks.

What would be most tragic about such losses is that actions needed to slow global warming already are well known: Reduce emissions of CO_2 and other greenhouse gases from fossil fuel-burning power plants, factories and automobiles. Yet the world's industrial nations—most notably the United States, the planet's biggest polluter—have so far done very little to solve the problem. In a recently released report, *U.S. Climate Action 2002*, the Bush administration acknowledged that continued build up of greenhouse gases would damage wildlife and ecosystems, yet it did not endorse any meaningful steps to reduce pollution.

The report pleased neither industry spokesmen, who claimed its message was too dire, nor environmentalists. "The administration now admits that global warming will change America's most unique wild places and wildlife forever," says NWF President Mark Van Putten. "How can it acknowledge a disaster in the making and then refuse to solve the problem, especially when solutions are so clear?" ❖

Can Birds Adapt?

Early birds beware: Breed too soon, and the worms needed to feed hungry hatchlings may be nowhere to be found. Yet such loss of synchronization with food sources is exactly what many scientists fear will happen more frequently as birds migrate and breed earlier in the year in response to warming climate. Even with only moderate warming, some species are already arriving at breeding territories before food is available for their offspring.

Still unknown is the extent to which birds can learn from such mistakes, adjusting the timing of migration or reproduction the following year to better coincide with food abundance. Recent research published in Science provides encouraging hints that at least some species can make adjustments. In Heteren, the Netherlands, biologists at the Center for Terrestrial Ecology monitored breeding pairs of blue tits—small birds similar to chickadees—in outdoor nest boxes over a period of two years. Inexperienced at the start of the experiment, all birds tended to breed slightly later than the region's peak caterpillar population.

In the study's first year, scientists supplemented the diets of half the breeding pairs with caterpillars and mealworms. The following season, those pairs laid eggs on about the same date they had the year before. Birds that had not been given extra food, on the other hand, bred earlier the second year, and their eggs hatched closer to when caterpillars naturally were abundant. The results suggest that birds such as blue tits may be able to adapt to some of the ecological disruptions caused by global warming.

But Terry Root, a biologist at Stanford University, warns that "not all birds are going to exhibit such plastic breeding behavior." The least flexible species, in fact, may turn out to be those that also are most vulnerable. Worrisome support for that hypothesis comes from a recent study of red-cockaded woodpeckers in North Carolina. Analyzing nearly two decades of data on more than 200 pairs of these endangered birds, a team led by biologist Karin Schiegg of the University of Zurich found that females on the whole have been laying their eggs increasingly earlier in response to warming climate, as have many temperate zone species. Notable exceptions to that rule, however, were woodpecker pairs that included at least one inbred partner, which also produced fewer offspring than pairs with no inbred birds. Because imperiled animals often live in small, isolated populations that become inbred, "climate change poses a previously unknown threat that may hasten the decline of endangered species," concludes Schiegg.

Reprinted with permission of Les Line.

Questions

1. According to scientists, how much will average global temperatures rise by the year 2100?

2. How might global warming also affect a bird's habitat?

3. What already well-known actions are needed to slow global warming?

Answers are at the back of the book.

Environmental Issues Aspects and Solutions

26

Although society's concern for the environment and charitable contributions to environmental groups are at an all-time high, environmental problems seem to be escalating rather than dissipating. It is proposed that perhaps the tactics of environmentalists have been the problem. Goldstein suggests that environmentalists would do well to stop vilifying corporations, recognize that the economy is a factor in all environmental decisions, be willing to negotiate, and to stop exaggerating environmental concerns.

Too Green for Their Own Good?

Andrew Goldstein

Time, August 26, 2002 (Special Report: Green Century)

HERE'S A RIDDLE TO KEEP YOU UP AT NIGHT: How come, at a time when the environmental movement is stronger and richer than ever, our most pressing ecological problems just get worse? It's as though the planet has hit a Humpty-Dumpty moment in which unprecedented amounts of manpower and money are unable to put the world back together again. "Why are we losing so many battles?" wonders Gus Speth, dean of Yale's School of Forestry & Environmental Studies.

Of course, the issues are complicated and could require decades and trillions of dollars to resolve. But part of the problem is that it's easier to protest, to hurl venom at practices you don't like, than to find new ways to do business and create change. The dogma of traditional green activism—that business (and economic growth) is the enemy, that financial markets can't be trusted, that compromise means failure—has done little to save the planet. Which means it's fair to ask the question: Have some of the greens' tactics actually made things worse?

This is not to say there hasn't been progress since the environmental movement began. The air and water in the developed nations of the West are, by most measures, the cleanest they have been for decades, and the amount of land protected as national parks and preserves has quadrupled worldwide since 1970. But despite a record flow of financial resources (donations to U.S. environmental groups alone have risen 50% in the past five years, to more than $6.4 billion in 2001, according to the American Association of Fundraising Counsel Trust for Philanthropy), the planet's most serious challenges—global warming, loss of biodiversity, marine depletion—remain as intractable as ever, making environmentalists vulnerable to charges that green groups have prospered while the earth has not.

So it's time to look at the past tactics of many green groups and identify lessons to be learned.

Business Is Not the Enemy

Thanks to scandals on Wall Street, environmentalists who have been bashing "evil" corporations for years have suddenly found themselves with plenty of allies. But the planet needs profitable, innovative businesses even more than it needs environmentalists. "It is companies, not advocacy groups, that will create the technologies needed to save the environment," says Jonathan Wootliff, a former Greenpeace executive turned business consultant.

So how to turn corporations into partners in preservation? For starters, when companies make efforts to turn green, environmentalists shouldn't jump down their throats the minute they see any backsliding. Wootliff says he was exasperated to watch so many environmental groups take special aim at Ford Motor, arguably Detroit's most environmentally friendly carmaker, during the latest fight in Congress over fuel-efficiency standards (in which Ford, GM and Chrysler all fought to preserve the status quo). "For goodness' sake, stop alienating your supporters," he warns. "Going after Ford will mean fewer, not more, CEOs will turn around and say protecting the environment is the right thing to do."

When conservation purity is the only acceptable option, the biggest polluters will have no incentive to clean up their acts. Says Dwight Evans, executive vice president of Southern Co., a major U.S. energy producer: "If tomorrow we announced we were shutting down 25% of our plants to put in new, high-tech scrubbing devices, the headline would be, WHY NOT THE OTHER 75%? We don't get credit for what we've done, or for what we're going to do."

This is not to suggest that environmentalists should be spineless. The threat of boycott prompted Home Depot to promise to phase out its selling of wood from old-growth forests. The good news is that once an industry leader turns green, the rest often follow, fearful that consumers will punish them if they don't. Today every major home-improvement retailer makes an effort to sell only products certified to have come from sustainably managed forests.

Embrace the Market

There is a simple economic explanation for why many of China's cities have become shrouded in gray clouds of dust: it's cheap to pollute. Millions of Chinese drive mopeds and old automobiles that don't have catalytic converters, and much of the nation's electricity comes from coal-fired power plants. Technology from the 1950s, after all, is at bargain-basement prices. But that's because the prices don't reflect the hidden costs of air pollution: deaths from lung illnesses and millions of dollars wasted on health-care bills and lost worker productivity. The situation is the same the world over. The price of goods and services rarely reflects environmental costs.

A concerted effort to correct this basic flaw in the market could have a bigger payoff for the environment than would a thousand new national parks. But many environmental groups continue to oppose market-based environmental reforms and instead remain wedded to the "mandate, regulate and litigate" model of the past.

Take, for example, power-plant emissions in the U.S., which environmentalists blame for much of global warming.

In the mid-1990s, the Clinton Administration was fairly close to striking a deal with the power industry that would have established a comprehensive emissions-trading program. To gain some certainty for their long-range planning, the utilities would agree to mandatory caps on emissions that included not just nitrogen oxides, sulfur dioxide and mercury but also carbon. Companies would have the flexibility of meeting targets in the most efficient manner by buying and selling emissions rights.

This didn't suit many of the environmental groups involved in the negotiations that believed the market was just a clever way for corporations to skirt environmental regulations. Says Katie McGinty, then chairwoman of Clinton's Council on Environmental Quality: "Practically every utility in the country began to accept the notion that they would face legally binding carbon restrictions. But environmentalists who were opposed to doing anything consensual with industry said what we really should be doing is suing their butts under the current provisions of the Clean Air Act." Result: today the U.S. Environmental Protection Agency has no ability to regulate carbon, and the old, pollution-spewing plants are still in operation.

It's Not All or Nothing

Toward the end of a war, a simple truism applies: it is better to negotiate a surrender than to fight to the death for a losing cause. Though environmentalists may be loath to admit it, this is their choice in the battle over genetically modified foods. Despite the best attempts by European activists to seal off the Continent from what they call Frankenfoods, the new science of farming is here to stay. So if environmentalists want to help shape the future of agriculture, it's time to raise the white flag and ask the world's bioengineers for a seat at the bargaining table.

What could be better for the environment than a cheap, simple way for farmers to double or triple their output while using fewer pesticides on less land? According to Rockefeller University environmental scientist Jesse Ausubel, if the world's average farmer achieved the yield of the average American maize grower, the planet could feed 10 billion people on just half the crop land in use today. Of course it's possible that some genetically modified foods may carry health risks to humans (although none have so far been proved in foods that have been brought to market), and it's unclear whether agricultural companies will be able to control where their altered-gene products end up. But what's needed now are not crop tramplers and lab burners but powerful lobbyists able to negotiate for more effective safeguards and a greater humanitarian use of the technology.

Bioengineering has tremendous potential in the developing world. The U.S., Canada, China and Argentina contain 99% of the global area of genetically modified crops, whereas yields of sorghum and millet in sub-Saharan Africa have not increased since the 1960s. Green groups hoping to earn the trust of the developing world should lobby hard for the resources of Big Agriculture to be plowed into discovering crop varieties that can handle drought and thrive on small-scale farms.

No More Exaggerations

A shattering piece of news came over the press wire of the Rainforest Action Network in May: "One-quarter of mammals will soon be extinct." An Associated Press story made a similar claim: "A quarter of the world's mammal species—from tigers to rhinos—could face extinction within 30 years." Problem is, the story isn't true.

The source of the number was a report issued by the United Nations Environment Program. It cites the World Conservation Union's most recent "Red List," which indicates that about 24% of mammals "are currently regarded as globally threatened." This figure comprises not only the approximately 4% of mammals that are "critically endangered" but also those that are merely "vulnerable," a category including animals with only a 10% chance of extinction within 100 years. The U.N. report makes this distinction clear—and even cautions against relying on species data from the Red List. But those caveats didn't make the news.

Fuzzy math and scare tactics might help green groups raise money, but when they, abetted by an environmentally friendly media, overplay their hand, it invites scathing critiques like that of Danish statistician Bjorn Lomborg, whose book The Skeptical Environmentalist debunks environmental exaggerations (see box).

Even more dangerous, notes Don Melnick, head of the Center for Environmental Research and Conservation at Columbia University, is how doomsayers create a Chicken Little problem. "We need to bury the notion that the biological world is going to collapse and we're all going to be extinct," he says. "That's nonsense, and it can make people feel the situation is hopeless. We can't have people asking 'So why should we bother?'" ❖

Questions

1. What are some of the planet's most serious challenges?
2. What is the good news about an industry leader turning green?

3. What does the World Conservation Union's most recent "Red List" indicate?

Answers are at the back of the book.

27

Because the mass media plays such a vital role in socially constructing the importance of environmental issues and disseminating information regarding solutions to such issues, it is of the utmost important that the linkages between the media and environmental activism are fully explored. An examination of the past and potential effects of the way in which natural processes, environmental degradation, and solutions are depicted in documentary filmmaking follows.

Seeing Green: Knowing and Saving the Environment on Film

Luis A. Vivanco

American Anthropologist, December 2002

IT IS CLEAR THAT THE MASS MEDIA PLAYS a crucial role in the construction and communication of environmental problems and solutions. There can be little doubt, for example, that the rise of tropical deforestation as an international political problem and site of transnational civic activism is due in no small part to the circulation of mass media representations, especially in the North (O'Connor 1997; Stepan 2001). It is also apparent that not all media are created and consumed equally. Knowledge of environmental problems relies heavily on nonmediagenic mechanisms of claims making and expert knowledge, and so the mass media often features spectacular and dramatic visual events or processes to draw attention to such claims (Hansen 1993). In other words, we can be told that rain forests are disappearing at alarming rates, but the often-cited statistics are not as compelling until one has actually seen the destruction. For example, because they "show" dramatically on film, environmental problems in a place like Costa Rica (the site of my ethnographic research) are often communicated mainly as the loss of wilderness and species through deforestation, at the expense of less visually appealing activities like pesticide abuses resulting from inadequate regulation of multinational

agricultural industries like bananas (Hilje et al. 1987; Thrupp 1990; Vivanco 1999).

Given the complex role of environmental and media bureaucracies in defining and communicating environmental problems (Chapman et al. 1997; Wapner 1996), the linkages between media representations and environmental activism are difficult to measure and easy to overstate. Such connections do provoke interesting questions, though, specifically around how "the environment" is represented visually, and the destruction and conservation of nature in the cinematic imagination. What are the various ways that natural processes are represented on film, and what do they tell us about the compatibility of nature and film? What are the problems posed by depicting environmental degradation and solutions to it on film? How can documentary film, with its conventional stylistic commitment toward realism and objectivity, represent environmental issues in cultural contexts in which ways of knowing and representing nature and social change can be quite distinct from those of Westerners?

This essay considers how such questions are currently being treated in films and film studies whose main concern is in a broad sense "environmental." The representation of nature as a conservation and management problem is intertwined with the invention, development, and commodification of cinematic technologies,[1] and environmental films offer a privileged site to consider how nature and its problems are visualized at the crossroads of science, popular education, art, and business. Yet there is no unified genre of "environmental films" with its own codes, structures, or even topics, which can range from the dangers of nuclear energy in industrial societies to wild animals in their habitats. My emphasis here is mainly on documentary works from the past five years so that I may provide information on recent releases of potential interest to anthropologists. What these works have in common is that they are "stories about stories about nature" (Cronon 1992), blending carefully crafted scientific and educational perspectives (and pretensions) with normative visions of the causes and consequences of environmental degradation. In their efforts to draw attention to potential solutions, they also play on (and sometimes challenge) iconic differentiations and conflicts between peoples, portraying morality tales of Third World peoples living in an either destructive or harmonious relationship with nature, and of heroic environmental activists struggling against forces of history. In the visions and arguments they articulate, these films provide useful grounds to consider what Escobar has called "the irruption of the biological" (1997:41), or the global emergence of nature's survival as a central political, ethical, and capitalist problem, and the role of the environment as an increasingly significant site of cultural production and catalyst of conflict, dialogue, and alternatives around the world.

The Fantasies of Natural History Film

It is widely accepted that wildlife and natural history films are environmentally committed documentaries (Bouse 2000). Two recent complementary books on the subject— *Wildlife Films* (2000), by cinema historian Derek Bouse, and *Reel Nature: America's Romance with Wildlife on Film* (1999), by historian of science Gregg Mitman-challenge this notion, asserting that the portrayal of nature on film and television generally takes shape as fictionalized narratives produced in a highly competitive corporate media marketplace. In fact, both reject the notion that these are documentaries at all, showing that since their origins in the early 20th century, these films have adopted devices, themes, and thrills of mainstream entertainment to ensure commercial success. Bouse, who calls them "docu-dramas," observes,

> For whether in two minutes or two hours, in a promotional trailer or a detailed natural history study, in nonnarrative montages or in elaborately plotted dramatic stories, wildlife film and television depict nature close-up, speeded-

up, and set to music, with reality's most enticing moments highlighted, and its "boring" bits cut out. [2000:2-3]

A central theme for both authors is the constitutive tension in natural history films between the fakery of simulated spectacle and the objectivity of science. Consider the recent public furor over revelations that Marty Stouffer, one of America's preeminent wildlife filmmakers, abused animals, staged scenes, and defiled public lands in the filming of his "Wild America" series. What this revealed, argues Mitman, "is how firmly Americans wish and expect nature films to be the real thing" (1999:204), based on unadulterated access to authentic and unspoiled landscapes and animals. Filmmakers themselves acknowledge that their products are somewhere between, as Bouse suggests, representation and simulation. He quotes cameraman Stephen Mills:

> When we film lions gorging on a bloody zebra in the Serengeti, or a cheetah flat out after a bounding gazelle, we rarely turn the cameras on the dozen or so Hiace vans and land-rovers, packed with tourists sharing the wilderness experience. All over the world we frame our pictures as carefully as the directors of costume dramas, to exclude telegraph poles and electricity pylons, cars, roads, and people. No such vestige of reality may impinge on the period-piece fantasy of the natural world we wish to purvey. [Bouse 2000:14]

The fantasy is reinforced by other techniques of camera placement and editing, such as concealing the camera and using telephoto lenses to capture "natural behaviors," using artificial lighting, and applying stock sounds during editing (Mitman 1999:24).

Just as important is the dramatic framework around which films are organized. The archetypal narrative structure of so-called blue-chip (classic) wildlife films is biographical and individualistic, tracing the individual from birth, through the perils of youth, trials of adolescence, adulthood, and death (Mitman 1999:131). Narratives about animal relationships often focus either on Darwinian "survival of the fittest" themes that emphasize the inevitability of individual acts of aggression and violence (what Andrew Ross [1994] calls the "Chicago Gangster Theory of Life") or, on animal kinship relations, in which culturally preferred notions of monogamy, responsible parenting, industrious work ethic, deferred gratification, and the sexual division of labor are presented. Especially violent and sexualized scenes deemed excessive for broadcast television audiences are self-censored, and sometimes sold separately as "outtakes." In effect, these films place a mirror in the face of Americans and-surprise!— they find their dominant cultural assumptions and ideals reflected in nature, even though the fields of reference the films create are a world apart from viewers' lives (Price 1999).

Within the industry, where films are as much commodities as educational tools, the self-conscious dilemma is that if filmmakers reveal these strategies, audiences who participate in the fantasy of wild nature will be disillusioned.[2]

The blurring between showmanship and science reflects the fluid boundaries between the uses of film as a research and educational tool within the sciences and the rise of film as a technology of mass communication and entertainment during the past century (Mitman 1999:60). Like early ethnographic films, nature films have been promoted as scientific records that preserve a natural heritage being lost to the forces of modernity (Mitman 1999:27). Scientists and filmmakers self-consciously placed limits on showmanship early on, as seen in the professional condemnation of Martin and Osa Johnson, who often staged nature as a spectacle of death and struggle in their films. Ironically, this has also meant that the scientific validity of films relies on perceptions of the filmmaker's virtue and integrity. Ethologist Konrad Lorenz, for example, brought his integrity to the well-respected *Encyclopedia Cinematographica* (Wolf 1952), whose goal was to document single "typical" behaviors of individual species usually around acts of communication. As Bouse argues, this technique promoted a showman's tendency to distill the natural world into a series of dramatic movements on film (he calls it "canned behavior"), later conventionalized in wildlife films through close-up shots that simultaneously decontextualize acts and promote feelings of intimacy and anthropomorphism (2000:72).

Both authors express gravitas when they consider the implications of these films for environmental change. Mitman is worried that they draw from and reinforce a gap between reality and expectation, whose root lies in a deep incompatibility between nature and the commercial impetus of the film industry. He observes, "Nature is not all action.... Conditioned by nature on screen, we may fail to develop the patience, perseverance, and passion required to participate in the natural world with all its mundanity as well as splendor" (Mitman 1999:207). Referring to the biographical structure and culturally specific themes that consistently show up in films, Bouse worries, "When moral values are presumed applicable to nature, their universalizing is complete. They become absolute, no longer moral values at all, but moral truths" (2000:159). In the end, perhaps because his work is more focused on the technical and institutional aspects of how filmmakers create nature as wild, Bouse is more cautious than Mitman about the presumed linkages between television viewing and the environmental movement.

These books provide crucial perspectives on the historical and institutional contexts in which dominant cinematic images of nature are reproduced and distributed, but there is a frustrating lack of attention to the diverse audiences that watch natural history films. Mitman's America is mainly white and middle class, and his conclusions about the meanings of nature films for viewers are more speculative than tested. More importantly, given the circulation of such films and their images beyond the United States, one wonders about viewer reception in the places that are portrayed as unpeopled (think East African savannas or rain forests), where there are ongoing struggles over access to resources desired as wilderness reserves.

Framing Environmental Degradation

Processes of environmental degradation carry an aura of crisis, usually confirmed on film by images of destruction—in-progress. Natural history films, for example, communicate an animal's prospects of individual (and group) survival by inserting cut-and-paste stock images of habitat destruction like logging.[3] Yet relatively few films dedicate themselves solely to the theme, perhaps because audiences do not respond well to pessimistic "doom and gloom" scenarios (Bouse 2000:16). The tendency in most is to briefly establish the problem on their way to examine solutions, reflecting an ideology of constructive engagement based on action (Grierson 1979). Lost in such a strategy can be an understanding of the differentiated and heterogeneous causes, consequences, and experiences of environmental degradation. Any number of examples could serve to illustrate this point, but consider the recent film *On Nature's Terms* (de Graaf 2001), which explores the challenges to large predatory mammals in the United States. The film argues that wolves, pumas, and bears are "keystone species," or important indicators of healthy ecosystems, because they require so much open space and regulate other animal populations. Prejudicial attitudes, government policies, and suburban sprawl, however, have contributed to their extermination, with consequential impacts on ecosystem health. That is more or less the extent of it; the film then presents how citizen action is conserving lands and contributing to species preservation. In arguing for the effectiveness of citizen action, it provides no detail on the contours of Americans' changing attitudes toward predators and predator loss, much less the diverse political and social meanings people make of arguments about predators and their roles in ecosystems.

Presentations on degradation often take shape through discourses of blame in service to a specific solution (Hoben 1995; Leach and Mearns 1996; Roe 1991), such as focusing on deforestation or poaching to justify the creation of absolute protection preserves. Two recent films claim both a fuller presentation on processes of degradation and confound the easy assignation of blame by focusing on the heterogeneous ways people experience and participate in environmental degradation. The first, *Since the Company Came* (Hawkins 2000), takes place on Rendova in the Solomon Islands, where a Malaysian logging company has been cutting and exporting tropical timber. The film's opening frames feature a man cutting down a tree with a chainsaw (a common scene to evoke larger destruction) and then cut to an emotionally charged meeting of the indigenous Haforai Development Corporation in which villagers are vigorously debating a decision to renew their contract with the Malaysian company. The film is organized loosely around this meeting from a fly-on-the-wall perspective and interspersed with villager interviews. Except for some brief intercuts of an early-20th-century silent film on the Solomons, the film never offers the views of non-Haforai (such as government officials, who at one point are accused of allowing the logging to continue, or of Malaysian logging company officials).

Especially without an off-screen voice-over narration, the film validates "local people" as important actors with a range of perspectives on logging and its impacts, in contrast to narratives that rely solely on experts to tell the story of degradation through objectivist language.[4] It also highlights social rifts and conflicts that accompany deforestation without falling into a common trap in environmentalist discourse on indigenous peoples: that environmental degradation leads to a unidimensional loss of culture. Gender politics is at the heart of this film's representation of conflict: While customary lands are inherited through matrilineage, men work as loggers and make decisions about logging contracts. The film depicts women who are especially critical of logging and resentful that their voices are ignored in decision making. According to one woman, "To us women, logging is destruction and pollution. We're affected by the pollution in the sea because we collect seaweed to feed our children.... All our fishing grounds now, after logging, we don't have enough fish" (Hawkins 2000) We also see the outlines of an emerging struggle over the formal conservation of the as-yet unlogged and uninhabited island of Tetepare, in which a young woman notifies the camera of plans to resist its future logging.

What has changed in Rendova, according to some of the interviewees, are attitudes toward land and the increased importance of the cash economy. One woman asserts, people tend to think they don't get much out of the land, when in fact they rely on land for gardens, for all they need. Food, water, everything come from land, but they don't see money coming out of it, so they don't think they benefit from their land or their forest. [Hawkins 2000]

Even some men who work as loggers are ambivalent about the industry and its effects, denying any easy divide between "pro-" and "anti-" logging perspectives. According to one logger, the devastation is obvious ("rivers that were once crystal clear and now they are muddy" [Hawkins 2000]); and he blames it on the government that since independence (1979) has pursued policies of export-driven resource extraction. Another logger observes that the problem is foreign domination itself: "Of course I cut down trees working as a logger, but I still think about the future too.... Instead of accepting big logging companies, Asian companies, we should have small mills and cut the timber ourselves" (Hawkins 2000).

Where this film falters is in providing a richer context for claims like those above. For example, during the meeting, the chief mentions a history of failed Haforai business enterprises, but we have no context for Haforai attitudes toward and experiences with "business" and "development" (Watson-Gegeo and White 1990). Instead of providing interviews that elucidate these themes, the film juxtaposes its present with old clips from a newsreel that celebrated the positive impacts of Lever Brothers and the plantation economy in bringing civilization and productivity to the headhunters. The technique is common (Stocking 1992:398), and the filmmaker's suggestion is obvious: There is a historical continuity in which Solomon experiences with outside commerce and development are based on domination and the unilateral extraction of natural and labor resources. The title itself suggests that forces external to the island are responsible for local destruction and conflicts; but because this film is not about "the company" as a subject, it obscures the institutional, national, regional, and transnational linkages that help shape the context in which deforestation happens (Utting 1993). Unfortunately, this also reinforces a trope of "good tradition" and "bad present," in which conflict in the present is naturalized as the product of dehistoricized foreign domination.

A second film, *Second Nature: Building Forests in West Africa's Savannas* (Maughan 1996), is more attentive to place-specific categories and practices, focusing on how they differ from broader narratives of environmental degradation. Based on and featuring the research of anthropologists James Fairheard and Melissa Leach on the savanna-forest mosaic landscape of Kissidougou prefecture, Guinea, West Africa, *Second Nature* adds to recent anthropological critiques of environmentalist and development crisis narratives that uncritically place blame on local ecological mismanagement and overpopulation (Fairhead and Leach 1996; Hoben 1995; Leach and Mearns 1996; Roe 1991). The first frames are self-conscious images of burning trees, but the film quickly establishes that this burning is part of a careful regime of indigenous land management, not an ominous loss of biodiverse habitat.[5] The point is to illustrate a provocative Boserupian discovery: What colonial and development officials have consistently interpreted as a once-continuous forest fragmented by Malthusian population growth and native mismanagement is in fact a carefully constructed landscape of forest patches cultivated on grasslands, and the product of population growth itself. In the course of the film, forest stewards explain that the forests serve as protective fortresses around villages, as hunting grounds, to prevent grassland fires from reaching villages, and to improve grassland soils for cultivation. The point is that such elite misreadings of environmental history are not inconsequential: They underestimate the extent to which any landscape is constantly modified by human intentionality in ways that can even enhance biodiversity and risk imposing ill-designed solutions on rural peoples (such as population control programs, instead of, say, technical support for reforestation).

The reason to watch this film is not so much for interesting visual material, and it offers a rather conventional narrative of (in this case, anthropological) experts and locals explaining stages in a process (forest growth) that because of its historical basis is difficult to show. The film also suffers from the larger dilemmas of Fairhead and Leach's research, including ambiguity about whether or not farmers and scientists experience different objective realities and the methodological limits of using aerial photography to make their case of forest expansion over time (Cleveland and Powell 1998:579). In its critical scrutiny of the nature-culture divide, however, it does raise the insight that environmental degradation itself cannot be considered independently from specific ecological meanings, practices, and histories. More importantly, it lays effective groundwork for turning the lens on environmentalists and scientific authorities themselves.

Visualizing Green Crusades

Films about conservation efforts can have a formulaic david-and-goliath quality, offering compelling visual material like conflicts between poachers and rangers, popular protests, and heroic activists struggling against oppressive bureaucratic forces. They also, however, inevitably raise basic questions (acknowledged or not) about what strategies will work to "save the environment," the forms of knowledge on which environmental practices are based and the social orders and hierarchies such practices imagine, and the role of the individual in promoting structural change. Furthermore, they communicate in visual and narrative terms basic tensions about the differences between groups of people-especially their capacity to offer positive models of ecological wisdom-as well as normalizing assumptions about who defines appropriate relationships to nature in conservation initiatives.

A dominant narrative, as seen in National Geographic films like *Save the Panda*, set in China (Birch 1983), and *The Rhino War*, set in Kenya and Zimbabwe (Clayford 1987), focuses on the actions of morally enlightened individuals whose goal is to save a species from the ignorance, greed, and overpopulation of local people. These films draw heavily on techniques of natural history films, including images of wildlife set to evocative music and dramatic storylines of destruction and redemption, although they introduce a new element: the neocolonial figure of the adventurous and self-sacrificing white Westerner whose task is to objectively understand the animal and defend its conservation (a role played in Save the Panda by the famous conservation biologist George Schaller, and in *The Rhino War* by white excombatants in Zimbabwe's civil war). Both films portray rural people in insultingly simplistic ways: They are silent background figures, as in *Save the Panda*, where they have no relevant perspective on pandas and their main job is as porters or assistants for Western scientists; threatening and desperate poachers; or reconstructed allies of conservation who, especially in the case of *The Rhino War*, which is about the low-intensity warfare between poachers and park rangers, will put their and poachers' lives on the line to help save an animal for their country's national development.

These films confirm two basic points familiar to natural history film audiences: (1) that proper conservation is to set up parks managed by experts and patrolled by armed guards, supported by efforts to educate locals about natural resources while keeping them separate from each other; and (2) that individuals, with the appropriate knowledge (science) and equipment (radio transmitters and guns) can make a difference. The strong tendency to rely on racializing stereotypes of Otherness, that in Save the Panda reaches absurd heights with orientalizing music and vaguely threatening descriptive language when referring to modern (Communist) China, supports a worldview in which wilderness conservation is the pinnacle of Western modernity. This has important consequences, one being that the complex histories and ecological knowledge of local people are deemed irrelevant, if not viewed as obstacles to overcome. As the *The Rhino War* seems to conclude, perpetual mistrust and even militarization are the sadly inevitable but justified result. The other is to homogenize conservation strategies and knowledge, confirming (erroneously) that there are no debates over the effectiveness of, much less alternatives to, concepts of conservation oriented toward dispossession, a static landscape vision of absolute protectionism, and scientific interventionism.[6]

A common alternative to this overtly negative orientalism and scientific triumphalism is the theme of the "green primitive," the romantic notion of the ecological noble savage living in harmony with nature. The Penan of Sarawak, Malaysia, became international celebrities in this respect during the 1980s and 1990s when some of them resisted logging in their territories with the help of a Swiss adventurer named Bruno Manner. This story gained extensive media coverage, including BBC and National Geographic films, and an influential 1988 Bullfrog Films release called *Blowpipes and Bulldozers* (Kendell and Tait 1988). This history is revisited in the Swedish documentary *Tong Tana: The Lost Paradise* (Roed et al. 2001). While the older films have the aura of urgency, the newer film offers a retrospective lament on the paradise Manner found, lost, and then sought to regain when the filmmakers accompany him back to Penan as he returns for "the final struggle" against logging.

International attention to the Sarawak situation has often been based on an unapologetic romanticization of the Penan as pristine innocents and the remarkable ability of a courageous individual like Manner to help galvanize their resistance (Brosius 1999). The makers of *Tong Tana* admit that, like Manner, they have been seduced by a romantic vision of the jungle and the splendid isolation of the Penan community they have come to help. Even while they see the Penan through Manser's eyes, they seem to have doubts about adopting his full-scale search for timeless primitives and gemeinschaft, pointing out that "Bruno is more Penan than the Penan themselves" (Roed et al. 2001). They also hint that the Sarawak struggle is more complex than the "FernGully allegory" (Brosius 1999) of films like *Blowpipes and Bulldozers*, in interviews with the minister of environment and a German sustainable forestry official.[7] Both use bureaucratic language to criticize Manser's activism, though for different reasons: the government official, on sovereignty grounds (reflecting the claims of ecoimperialism the Malaysian government used to effectively challenge the international Sarawak campaigns), and the sustainable forestry official on the grounds that, while Manser did help generate international pressure, he missed opportunities for capacity building among the Penan. The film concludes ominously, with the notification that Manser is probably dead, self-consciously identifying the end of a hopeful era of resistance among the Penan and affirming the grim reality that tropical forest and cultural destruction continue unimpeded.

A central problem with green primitive allegories, though, and *Tong Tana* falls into this trap in spite of its own

sense that it should not, is that the overwhelming desire for purity denies the voices and agency of the very people it claims to represent. The Penan are central players in this drama, yet we never find out in depth about what they think about either Manser or their situation. They are presented in characteristically unidimensional terms, as either nomadic (pure) or settled (contaminated and losing their culture), confounding more complex realities and in the process assigning them to the natural (not historical) side of the nature-culture divide. Consequently, we remain ignorant of the divisions between Penan who allied with logging companies and Penan who allied with environmentalists as extensions of their own historical conflicts with one another, as Peter Brosius has shown in his compelling research on the Sarawak campaign and Penan activism (1997, 1999). By ranging between romantic and bureaucratic languages, *Tong Tana* ultimately reinforces a broader depoliticization of rain forest conflicts, eliminating the claims of those who do not conform to idealized images (Conklin 1997:728) while validating interventions defined within bureaucratic institutions (Brosius 1999).

As Conklin has pointed out, representations and self—presentations of indigenous authenticity can have their rewards in ecopolitics. Exotic peoples in traditional dress and body decoration offer keen visual material, marking indigenous difference and identity in ways that affirm their closeness to nature (1999:722-723). As documented in the work of Terence Turner (1992) and in the films *Kayapo: Out of the Forest* (Beckham 1989) and Geoffrey O'Connor's *Amazon Journal* (O'Connor 1995), Kayapo have gained international media celebrity by playing up the role of green primitive and noble savage in Western environmentalist discourse, which provided leverage that enabled them a new level of political opportunities and influence, including success persuading the World Bank to not loan the Brazilian government funds for the Altamira dam. At the same time, the spread of representational and communications technologies have created opportunities for indigenous groups to represent themselves in ways informed by their own aesthetics of social production and value, actively placing themselves at the intersection of cinematic imagery and ecopolitics. Nevertheless, as the Kayapo and other marginalized groups have discovered, there are trade-offs involved in aiming such symbolic politics at international audiences, including the dilemma of playing roles handed to them by dominant discourses that prohibit variation from idealized imagery. Furthermore, the success of international campaigns can become a liability at the national level where the claims of marginalized or minority peoples often represent a challenge to the legitimacy and hegemony of the modern states that encompass them (Conklin 1997:725; van den Berghe 1992).[8]

Surely, environmental activism is not a static or inflexible arena, and it is a highly complicated and even contradictory site of cultural politics and social production. In the era of "sustainable development," mainstream environmental institutions and funding agencies have begun to openly acknowledge that stark separations between people and nature, and alternatively simplistic divides between primitive peoples and modern society, which collapse indigenous societies into nature, do not necessarily lead to effective conservation initiatives. Evicting people from the landscape and disrupting traditional productive activities can disempower and impoverish the very communities conservation claims to benefit, thereby generating new threats to nature.[9] So there is a growing realization that conservation is not simply about what kind of nature activists imagine or know they want to preserve or restore; it is also an important arena in which they, explicitly or implicitly, project and reimagine community, political-economic relationships, and social justice.

Reflecting some of these shifts is a new film that has generated substantial popular and festival attention since its release in 2001. *The Shaman's Apprentice* (Smith 2001) focuses on the work of Dr. Mark Plotkin, a North American ethnobotanist who started an Amazonian conservation program called "The Shaman's Apprentice" that seeks to facilitate the transmission of ecological and medicinal knowledge from indigenous elders to young people in an effort to preserve and regenerate Amazonian cultures and secure biodiversity conservation. Narrated by Susan Sarandon, the film trades on the considerable public curiosity about shamanism, mixing techniques like dramatic reenactments, slow motion, archival images, sepia maps, images of sublime nature, and so on, with objectivist storytelling in a documentary format. Recounting episodes from his book *Tales of a Shaman's Apprentice*, Plotkin is the hero at the center of the film, demonstrating the subtle and encyclopedic knowledge of healers and discussing through interviews the lamentable fact that such knowledge is dying because young people are taught to revere Western medicine over shamanism. He tells us, "The most endangered species in the Amazon is the shaman himself.... When a shaman dies, it's like a library burning down."

Mixed metaphors aside-native peoples are naturalized as "endangered species" and civilized as "libraries"—this film substantiates claims of sophisticated indigenous intimacy with biodiversity that green primitive narratives only imply and other conservation films ignore. As self-conscious salvage ethnographer, Plotkin urges us to consider the empirical rigor of shamanistic biochemistry ("they're better chemists than us in certain instances"), and argues that Western medicine's materialistic bias and search for magic bullets prevents a holistic understanding of disease etiology and the subtle and comprehensive understanding of nature among indigenous healers. In a reenacted scene from the book (that also inspired a scene in Sean Connery's *The Medicine Man*), we see Plotkin's conversion from skeptic to believer in a dream episode where he is visited by a shaman, which demonstrates the importance of dreams to shamanistic ecological and medicinal knowledge. As a result, Plotkin self-consciously sets himself up as a scientist-hero: a renegade, sophisticated in the ways of the locals, and not afraid to critique the epistemological limitations of his own knowledge to get at the heart of the problem. Plotkin also knows that his audience (and funders) will include skeptics, and so he is careful to validate Western science's objective methodologies, including demonstrations of how a botanist prepares an herbarium and translations of indigenous cures into biochemical categories.

What remain inexcusably hazy in this film are issues concerning indigenous collective property and traditional resource rights, and the controversial politics and economics around the patenting of indigenous knowledge and plants in international trade regimes that tend to favor pharmaceutical industry interests. Plotkin admits that ethnobotany has been what he calls "a rape and run enterprise," although the reason for saying this is less to explore how and why it continues, than it is to set up his own conservation program as mutually beneficial solution for indigenous peoples (it perpetuates their culture) and ethnobotany (it learns a lot about cures for disease). At the end of the credits, a statement vaguely answers a question that haunts the rest of the film: "ACT [Plotkin's Amazonian Conservation Team] does not engage in bioprospecting" (Smith 2001). Amazingly, the film does not define, much less acknowledge, the rise of bioprospecting and biopiracy. It therefore does not place "The Shaman's Apprentice" program in the wider context of growing awareness of the problems associated with bioprospecting and the political-economic inequalities that underlie them. These include no or poor compensation to communities by the capitalist enterprises that process and market the products of bioprospecting, payments to individuals and not collectives, and outright illegal collection of plants and appropriation of knowledge (Shiva 1997). The film also ignores the galvanization of indigenous peoples around these problems, some of whom are forming strict collection guidelines, creating networks and centers that advocate for the protection of ecological and medicinal knowledge, and, in celebrated instances, internationally denouncing questionable practices and the abuses of bioprospecting, such as the recent situation in Chiapas in which University of Georgia ethnobotanists clashed with indigenous peoples organizations (and ultimately failed to get permission to conduct bioassays on collected plant materials) (Action Group on Erosion, Technology, and Concentration [ETC Group] 2001).

These debates and new movements point to a broader trend in which indigenous activists and communities have been acting in coordinated ways to produce and control representations of themselves and their relationships with their environments. This is especially the case in the arena of ecotourism, in which some indigenous peoples have been actively contesting the mainstream acceptance of ecotourism as a strategy for sustainable economic development, cultural preservation, and nature conservation. The UN's declaration of 2002 as the "International Year of Ecotourism" has intensified these organizing efforts, leading to several international meetings this year (in Thailand and Oaxaca, Mexico) and the organization of the Indigenous Tourism Network, in which indigenous representatives have vocalized their concerns about the consequences of ecotourism development on their cultural integrity, land rights, and self-determination. At the International Forum on Indigenous Tourism, held in Oaxaca during March 2002, for example, participants pointed to ecotourism as a rationale that governments and international agencies (including bioprospectors) use to gain access to or control over indigenous lands, as well as the unequal participation of indigenous representatives in planning projects. Other negative impacts include poorly distributed profits, greenwashing, the overuse of habitat by tourists, and outright destruction of habitat to create tourism infrastructure (Vivanco 2002, in press).

The alternative visions of tourism emerging out of gatherings like the one in Oaxaca are not based on the rejection of tourists and ecotourism per se, but the rejection of the industry's market-driven universalism. More importantly, they affirm the diverse and nonprescriptive ways indigenous communities can engage in tourism to support and communicate their goals of cultural and political selfdetermination. The experience of Yavesia, a small village in the Sierra Juarez of Oaxaca, is indicative of this active appropriation and redefinition of ecotourism. The community has resisted creating "attractions" simply to bring tourists (who are allowed to visit community-owned forest land with local guides) and view their involvement in tourism as a chance to share stories about their stewardship of the landscape. Conscious of the tendency of tourism to displace other productive activities, the project is integrated into other uses of the forest, including a springwater bottling plant and the collection of nontimber forest products. Decisions regarding the project and the distribution of its modest economic benefits are made in community assembly, and work is organized through tequio, a responsibility to voluntarily participate in building and maintaining communal infrastructure. Visitor movements and guidelines are coordinated with neighboring villages of Ixtlan and Ixtepeji, in order to ensure that no single niche or community gets overrun, and that certain sacred spaces, activities, rituals, and knowledge remain protected from the tourists' gaze. The point of the Yavesia example is that nature-centered tourism is increasingly viewed as one of several strategies for community survival, cultural regeneration, and the communication of indigenous narratives about their own cultural ecologies. Unlike Plotkin's representation of indigenous concerns for cultural survival or the ecotourism industry's uncritical promotion of "sustainable development," territorial and political claims for self-determination are often at the forefront of how indigenous peoples represent nature, its conservation, and its visual appreciation.

Conclusion

The Shaman's Apprentice, like many other environmental films and the conservation programs they portray, has popular appeal because it offers a carefully crafted win-win vision of conservation and sustainable development, in which local participation and cultural survival are joined with the far-sighted work of visionary scientist-activists. As conservationist propaganda, however, it raises more questions than it answers. For whom is the rain forest actually being saved? That is, how do appeals to universal human ideals obscure the narrow interests that benefit, and what kinds of problems does this create for the people who actually live there? Given the high stakes for national and international bureaucracies competing to legitimate their authority over landscapes, who has access and who governs access to the rain forest?

In such circumstances we have more to gain by scrutinizing the vehicles of representation, including the realisms they project and the dilemmas they omit, than by taking their messages and images as disinterested indications of "how nature works" and how to resolve its problems. Aside from their artistic and commercial impulses, environmental films are earnest political documents, through which filmmakers and their subjects communicate concerns about nature, and even more so, about issues not strictly about nature. These films express the emergence of the social orders and practices deemed necessary to ensure nature's survival. They also serve to validate and perpetuate conservation institutions and initiatives themselves. Looking at them in this way does not imply that nature as reality is irrelevant but recognizes how thoroughly conflicting goals and practices of visual representation mediate knowledge and the moral certainties of environmental problems as well as the social organization of their solutions.

Acknowledgments

Special thanks to Jeff Himpele, AA Visual Anthropology editor, for constructive editing advice and help securing video review copies.

Notes

1. Consider, for example, Teddy Roosevelt's commitment to both landscape conservation and his participation in the film *Roosevelt in Africa* (1910), which documented a hunting trip to Africa. According to Mitman (1999:13), the masculine endeavor of hunting was at the time considered a "therapeutic balm" for the debilitating effects of urban living and was supported by conservation efforts that would encourage people to spend time in the great outdoors.

2. In light of the popularity of so-called reality television and the rise of wildlife blooper films on channels like the Discovery Channel and Animal Planet, it is apparent that viewers are quite interested in "behind-the-scenes" scenarios (even if they are highly edited).

3. Because nature films rarely show the "everyday life" of an actual animal whose habitat is actually being destroyed, these images provide dramatic emphasis in a zero-sum allegory in which human economies expand at the expense of natural populations.

4. The classic film *Environment under Fire* (Dworkin et al. 1988) offers a typical narrative strategy by using a variety of talking-head experts (politicians, scientists, and activists) to explain the nature and scope of environmental degradation in Central America.

5. The image of fire is an especially common one in environmental discourse to suggest the urgency of degradation. Famous examples include *The Burning Season* (Revlon 1990), which is about Amazonian rubber tapper and union organizer Chico Mendes, who became an icon of 1980s grassroots environmental activism, or Shoumatoff's *The World Is Burning* (1990), which is also about Amazonian deforestation.

6. Alternatives focus mainly on restoring community-level resource management authority and practices. One of the widely cited examples comes from Zimbabwe itself. The CAMPFIRE (Communal Areas Management Program for Indigenous Resources) program sought to generate support for wildlife conservation by setting up resource cooperatives in which communities manage wildlife populations and benefit directly from tourism or hunting licenses. CAMPFIRE has come under critical scrutiny because of conflicts between state bureaucracies, communities, and private initiatives (Derman 1995), and it has more or less disappeared with the political upheavals in Zimbabwe. Nevertheless, it has been widely referenced as an African alternative to the ejection of native peoples from the landscape (Adams and McShane 1992).

7. *FernGully: The Last Rain Forest* (Kroyer 1992) is a popular animated film that presents a story of timeless forest innocents faced with aggressive bulldozers and destruction.

8. Both of these themes are explored in O'Connor's film *Amazon Journal* (1995). The film, which is an extended critique of the green primitive ideals he has encountered in the Amazonian region and in international sustainable development conferences like Rio 1992, shows the access Kayapo gained in the international media (with the support of celebrities like Sting), as well media condemnation of a Kayapo leader when it was "discovered" that his lifestyle and pursuit of Western goods conflicted with "who Indians should be." The film also considers the marginalization of Kayapo within Brazilian politics, precisely because of their identification with international interests that conflict with nationalist and state intentions in the Amazon.

9. This is the most basic ideology of "sustainable development" as elaborated in the UN's Agenda 21, which boils down to the following: If human beings living in areas of rich biological diversity cannot live economically viable lives, they will inevitably destroy the resource base that keeps wildlife alive and kill the animals themselves. [Wapner 1996:83]

References Cited (Texts)

Action Group on Erosion, Technology, and Concentration. 2001 *U.S. Government's $2.5 Million Biopiracy Project in Mexico Cancelled*. Electronic document, http://www.rafi.org/article. asp?newsid=279, accessed June 23,2002.

Adams, Jonathan, and Thomas McShane. 1992 *The Myth of Wild Africa*: Conservation without Illusion. Berkeley: University of California Press. Bouse, Derek 2000 Wildlife Films. Philadelphia: University of Pennsylvania Press.

Brosius, J. Peter. 1997 Prior Transcripts, *Divergent Paths: Resistance and Acquiescence to Logging in Sarawak, East Malaysia*. Comparative Studies in History and Society 39:468–510.

1999 *Green Dots, Pink Hearts: Displacing Politics from the Malaysian Rain Forest.* American Anthropologist 101 (1):36–57. Chapman, Graham, Keval Kumar, Caroline Fraser, and Ivor Gaber

1997 *Environmentalism and the Mass Media: The North-South Divide.* London: Routledge.

Cleveland, David, and Joseph Powell. 1998 *Review of Misreading the African Landscape: Society and Ecology in a Forest-Savanna Mosaic.* American Anthropologist 100(2):578–589.

Conklin, Beth. 1997 *Body Paint, Feathers, and VCRs: Aesthetics and Authenticity in Amazonian Activism.* American Ethnologist 24(4):711–737.

Cronon, William. 1992 *A Place for Stories*: Nature, History, and Narrative. Journal of American History 78 (March 1992):1347–13 76.

Derman, Bill. 1995 *Environmental NGOs, Dispossession, and the State: The Ideology and Praxis of African Nature and Development.* Human Ecology 23(2):199–215.

Escobar, Arturo. 1997 *Cultural Politics and Biological Diversity: State, Capital, and Social Movements in the Pacific Coast of Colombia.* In Between Resistance and Revolution: Cultural Politics and Social Protest, Richard Fox and Orin Stam, eds. Pp. 40–64. New Brunswick: Rutgers University Press.

Fairhead, James, and Melissa Leach. 1996 *Misreading the African Landscape: Society and Ecology in a Forest-Savanna Mosaic.* Cambridge: Cambridge University Press.

Grierson, John. 1979 *Grierson on Documentary*. Rev. edition. Forsyth Hardy, ed. Boston: Faber and Faber.

Hansen, Anders, ed. 1993 *The Mass Media and Environmental Issues.* Leicester: Leicester University Press.

Hilje, Luko, Luisa Castillo, Lori Ann Thrupp, and Ineke Wesseling. 1987 *El Uso de Plaguicidas en Costa Rica.* San Jose, Costa Rica: Editorial UNED.

Hoben, Alan. 1995 *Paradigms and Politics: The Cultural Construction of Environmental Policy in Ethiopia.* World Development 23(6):1007–1021.

Leach, Melissa, and Robin Mearns. 1996 *The Lie of the Land: Challenging Received Wisdom on the African Environment.* Oxford: International African Institute, James Currey, and Heinemann.

Mitman, Greg. 1999 *Reel Nature: America's Romance with Wildlife on Film.* Cambridge, MA: Harvard University Press.

O'Connor, Geoffrey. 1997 *Amazon Journal: Dispatches from a Vanishing Frontier.* New York: Dutton.

Price, Jennifer. 1999 *Flight Maps: Adventures with Nature in Modem America.* New York: Basic Books.

Revkin, Andrew. 1990 *The Burning Season.* New York: Plume Books. Roe, Emery

1991 *Development Narratives, or Making the Best of Blueprint Development.* World Development 19(4):287–300.

Ross, Andrew. 1994 *The Chicago Gangster Theory of Life.* London: Verso.

Shiva, Vandana. 1997 *Biopiracy: The Plunder of Nature and Knowledge.* Boston: South End Press.

Shoumatoff, Alex. 1990 *The World Is Burning: Murder in the Rainforest.* Boston: Little, Brown.

Stepan, Nancy. 2001 *Picturing Tropical Nature.* Ithaca: Cornell University Press. Stocking, George

1992 *From Spencer to E-P: Eyewitnessing the Progress of Fieldwork.* American Anthropologist 94:398–400.

Thrupp, Lori Ann. 1990 Environmental Initiatives in Costa Rica: A Political Ecology Perspective. Society and Natural Resources 3:243–256.

Turner, Terence. 1992 *Representing, Resisting, and Rethinking: Historical Transformations of Kayapo and Anthropological Consciousness.* In Colonial Situations. George Stocking, ed. Pp. 285–313. Madison: University of Wisconsin Press.

Utting, Peter. 1993 *Trees, People, and Power: Social Dimensions of Deforestation and Forest.* Protection in Central America. London: Earthscan.

van den Berghe, Pierre. 1992 *The Modem State: Nation-Builder or Nation-Killer?* International Journal of Group Tensions 22(3):191–208.

Vivanco, Luis. 1999 *Green Mountains, Greening People: Encountering Environmentalism in Monte Verde, Costa Rica.* Ph.D. dissertation, Department of Anthropology, Princeton University.

2002 *The International Year of Ecotourism in an Age of Uncertainty.* Clearinghouse for Reviewing Ecotourism, vol. 23, available from the Tourism and Investigation Monitoring Team, tim-team@access.inet.co.th.

In press *Indigenous Struggles for Tourism Alternatives.* Alternatives Journal.

Wapner, Paul. 1996 *Environmental Activism and World Civic Politics.* Albany: State University of New York Press.

Watson-Gegeo, Karen Ann, and Geoffrey White. 1990 *Disentangling Conflict Discourse in Pacific Societies.* Stanford: Stanford University Press.

References Cited (Films)

Beckham, Michael, dir. 1989 *Kayapo: Out of the Forest.* 52 mins. Chicago: Films Incorporated Video.

Birch, Miriam, prod. 1983 *Save the Panda.* 60 mins. Stamford, CT: Vestron Video, with National Geographic Society and WQED Pittsburgh.

Clayford, Phillip, prod. 1987 *The Rhino War*. 60 mins. Stamford, CT: Vestron Video, with National Geographic Society.

de Graff, John, prod. and dir. 2001 *On Nature's Terms*. 25 mins. Okley, PA: Bullfrog Films.

Dworkin, Mark, Joshua Karliner, and Daniel Faber, dirs. 1988 *Environment under Fire*: Ecology and Politics in Central America. 27 mins. San Francisco, CA: EPOCA, Earth Island Institute and Seattle, WA: Moving Images Video Project.

Hawkins, Russell, dir. and prod. 2000 *Since the Company Came*. 52 mins. Brooklyn, NY: First Run/Icarus Films.

Kendell, Jeni, and Paul Tait, dirs. 1988 *Blowpipes and Bulldozers*. 60 mins. Okley, PA: Bullfrog Films.

Kroyer, Bill, dir. 1992 *FernGully: The Last Rainforest*. 72 mins. 20th Century Fox Home Entertainment.

Maughan, Graham, dir. 1996 *Second Nature: Building Forests in West Africa's Savannas*. 40 mins. West Sussex, UK: Cyrus Productions.

O'Connor, Geoffrey, prod. and dir. 1995 *Amazon Journal*. 58 mins. New York: Filmmakers Library. Roed, Jan, Eric Pauser, and Bjorn Cederberg, dirs.

2001 Tong Tana: The Lost Paradise. 52 mins. New York: Filmakers Library.

Smith, Miranda, dir. 2001 *The Shaman's Apprentice*. 54 mins. Okley, PA: Bullfrog Films.

Wolf, Gotthard, prod. 1952 *Encyclopedia Cinematographica*. Goettingen: Institute for Scientific Films. ❖

Questions

1. What does the film, On Nature's Terms, argue about "keystone species"? What does the film omit?
2. What basic questions do films about conservation efforts raise?
3. Ecotourism serves as a strategy for what?

Answers are at the back of the book.

Ecuador's marketing cooperative, the Callari Project, is proving to be a viable option for poverty-stricken indigenous rainforest populations. In the past two years, almost 1,000 artisans and farmers have managed to increase their income without harming the rainforest around them. Without the Callari Project, many people would be forced to either level the rainforest they call home or engage in dangerous trade with guerillas fighting in Colombia's civil war.

28

A Forest Path Out of Poverty

Arie Farnam

The Christian Science Monitor, August 9, 2002

DAVID ANDI, HIS WIFE, AND FOUR CHILDREN live in a bamboo hut in the Amazon rainforest, several hours' hike from the nearest dirt road. They grow corn and cocoa, scraping a living out of a tiny field cleared in the jungle. Ecuador's economic crisis in the late 1990s eroded the family's income to less than $500 per year, while prices for necessities tripled.

Until recently, Mr. Andi, one of the Quechua people, knew of only two ways to escape this poverty, and he liked neither: He could log the rain forest, which would leave landscapes of infertile soil; or he could trade with warring factions from neighboring Colombia and thereby abet the Colombian civil war's spread south into Ecuador.

Now he has an alternative. Seven months ago, he joined the Callari Project, a marketing cooperative spanning 15 towns and villages in Ecuador's Napo province. Callari, whose name means "ancient" in the Quechua language, aims to help indigenous people make a living without destroying their forest or getting involved in the Colombian conflict.

Since the cooperative formed two years ago, 700 artisans and 300 farmers have increased their incomes 30 percent by improving the quality of the cocoa and coffee they grow, relearning indigenous methods for producing useful items from jungle materials, and marketing their products abroad. Together they have made some $200,000 in extra income that goes to buy schoolbooks, build better houses, or provide emergency funds.

"My family's situation has improved a great deal," Andi says, while he weaves a mesh of strong fibers that he pulled out of the long leaves of a pita plant. "Now, we can buy school clothes and books for my children. They never could have those things before." Outside the capital, Quito, education is not free beyond elementary school. It costs $300 per year to send a child to high school.

Peace Dividend

But the effects of this project go beyond education. Thalia Flores, editor of the national newspaper Hoy,says Callari could provide a new economic model in the Amazon. "We desperately need economic alternatives like this in the Amazon region," she says. "This is the key to protecting Ecuador from the Colombian conflict. When people are so poor they have nothing left to lose, they will do anything to make money, even dangerous things. The Callari cooperative won't solve everything over night, but it can improve the security situation and provide a model for sustainable development."

Bartolo Tapui, a former policeman in Andi's village, says that before the Callari cooperative expanded to Ila Yaku, the Revolutionary Armed Forces of Colombia (FARC), the largest guerrilla group in Colombia, was gaining a foothold in the area. But as soon as the villagers organized, the guerrillas went elsewhere. "The FARC uses people who are desperate and weak," Mr. Tapui says. "Now that people here can make money doing honest work, they won't have anything to do with the guerrillas."

The cooperative was not formed to resist the Colombian combatants, however. It was initiated by a Kansan named Judy Logback, who came to Ecuador to do environmental education in 1997. She got to know indigenous communities and tried to persuade them to stop cutting down their trees for sale.

Seeds of Success

"One family explained to me that for the price of a hardwood tree, they could send one of their children to school," Ms. Logback says. "They know that it is damaging to cut down the trees, but when it is between a child's education and a tree, their choice is clear. A local teacher said to me, 'If you want to save the rainforest, help us find a way to make a living without destroying it.'"

She started by selling seeds the villagers collected to reforestation programs. That way, they could make just as much money from a living tree in a year, as they could by cutting it down. Logback also noticed the beautiful baskets, bowls, canoes, weavings, jewelry, and even toothbrushes that the indigenous people made from the forest's natural, biodegradable materials. By 1999, Logback was selling some of these crafts to tourists. The artisans, who were used to intermediaries who bought their necklaces for $1 and later sold them for a high profit, were amazed when Logback gave them the full $5 or 6 she was paid for the crafts. "Immediately everyone wanted to make 10 more," Logback says. "So, we had a problem. We needed a marketing strategy."

Logback, who received a degree in microbiology from Beloit College in Wisconsin, knew little about marketing, but her enthusiasm made up for it. Within two years she had developed a network of shops, museums, and friends in 10 countries that agreed to sell Callari crafts. Last year, the project received a grant of $240,000 from the US government rural aid program known as PL480, and the Canadian International Development Agency. The funds helped to broaden marketing efforts and hire community elders as teachers to improve on the quality of the crafts and farm produce. Last month, Logback was awarded the New York-based Bay and Paul foundations' Biodiversity Leadership award, which comes with a $180,000 grant.

Callari has quickly gained a reputation in Ecuador for doing the most with the least amount of money. "The small amount of money we gave the Callari project was only meant to test the idea," says Luis Sanchez, Ecuador director of PL480. "We did not expect any concrete results yet, and we were pleasantly surprised. This project has the highest results per dollar spent."

In the past three months alone, it has doubled the price to 50 cents per pound that farmers can get for their cocoa. The cooperative organized farmers to pool their money to rent a truck that would take their produce to seaports for sale, bypassing dishonest middlemen.

Logback says she hopes that the project will spread to other parts of Ecuador. She plans to make Callari a brand name with strict quality control. "I figure, if Coca-Cola can spread to the ends of the earth, so can Callari," she declares, as she sends off a shipment of crafts to Germany. ❖

Questions

1. What is the Callari Project?
2. Since the Callari Project formed, how have 700 artisans and 300 farmers increased their incomes by 30%?

3. Who initiated the Callari Project?

Answers are at the back of the book.

29

Environmentalists and farmers have traditionally been at odds over goals for environmental policies. Recently, however, these two groups have joined forces to combat common concerns such as concentrated animal feeding operations (CAFOs). In spite of traditionally divergent philosophies, these one-time enemies are beginning to see eye to eye on some issues.

Growers and Greens Unite

Gerald Haslam

Sierra, January/February 2003

BRIAN BLAIN IS A STRAIGHT TALKER who will tell you what you need to hear, which is not necessarily what you want to hear. "A lot of farmers genuinely believe that environmentalists are out to destroy agriculture," the head of California's largest pecan-growing and -processing operation announces. "Many things can be done that would be good for farmers and for the environment, but suspicion prevents people from working together." In the central-valley city of Visalia, Blain and the Sierra Club's Richard Garcia smashed the stereotypes when they took on a local irrigation district hell-bent on cementing an earthen canal and ruining its riparian habitat, home to century-old valley oaks and the San Joaquin kit fox. "We had to work together," explains Garcia. "The stakes were simply too high to let ourselves lose."

Common concerns can bring diverse groups to the table, but overcoming mistrust between farmers and conservationists has been particularly vexing. Environmentalists are often lumped in with outsiders—particularly the bureaucrats who dispense rules and regulations from afar—who don't understand what's really happening on the farm, and conservationists often discount farmers as stubborn and narrow-minded. Both characterizations have, at times, been apt. But

"grower-green" alliances are developing nationally, catalyzed by high-profile issues like the explosive growth of industrial livestock operations and the struggle to keep family farms afloat and maintain the rural character of local communities. (It's a good thing, too, since farms and ranches occupy more than half of all land in the Lower 48.) People with wildly different backgrounds are learning more about each other as they come together to defeat common opponents.

Concentrated animal feeding operations (CAFOs) are among both groups' most determined and well-funded foes. "There are hog farms here so huge they seem to extend from horizon to horizon, and their stench has unified farmers and environmentalists," explains Scott Dye, director of the Sierra Club's Water Sentinels Program. "The people who suffer as a result of hog factories aren't newcomers. They're people who've been farming for generations before the corporate hog operations showed up."

In Missouri, farmers defied the powerful Farm Bureau in 1998 and joined environmentalists to seek endangered designation for a small fish, the Topeka shiner. As cattle farmer Martha Stevens reasoned, "If the water kills the fish, it can't be good for us." That simple logic packed a hearing in the small town of Bethany, with local farmers concerned about what runoff from industrial hog farms was doing to their environment. "The shiner is an indicator that stream water is safe and clean," Stevens notes, explaining why small farmers supported the designation even though they would be required to keep soil out of waterways. "They figured it was something we could live with." While Stevens's water comes from wells rather than streams polluted by massive hog farms, she suspects the two sources are intertwined; besides, she says, "my kids used to wade and swim in these streams. They can't do that with their kids today. Not when the water is loaded with E. coli and nitrates and other crap."

According to Ken Midkiff, clean-water campaign director for the Sierra Club, environmental groups are in a good position to lend a hand on farm-related issues. "We deal with everything from organic to sustainable to pesticides to monoculture to CAFOs and everything in between," he says. "Farmers recognize that we are much more in tune with their interests than the commodity groups, particularly the Farm Bureau." With more than 5 million members, 2,800 county bureaus, and a Washington, D.C., lobbying staff, the American Farm Bureau Federation defines clout—and conservatism. Over the years, the organization has regularly opposed plans to benefit the environment, fearing even the slightest impact on agricultural profits. The Farm Bureau is critical of the Endangered Species Act, the Food Quality Protection Act (passed unanimously by Congress in 1996), and of attempts to regulate CAFOs. The thorn in its side is the 300,000-member National Farmers Union, which has allied with environmental groups to oppose farm subsidy payments that disproportionately benefit very large operations. Last year, the Sierra Club joined the union in a major lawsuit against corporate hog-raising factories in the South and Midwest. "Ag-enviro" alliances succeed by stressing what both sides have in common, a lesson not lost on Scott Dye. "Some farmers think a Sierra Club organizer will show up looking like their image of an environmentalist nut, with a long ponytail and sandals," he says. "When I show up looking like an average farm kid, well, they're more apt to listen."

Not that there isn't still skepticism. "Farm folks tend to be distrustful of city people," acknowledges Dye. "You have to go to the farm where a family has lived and worked for a hundred years, and listen to their concerns. You have to show genuine empathy, not phony sympathy. They can tell in a hurry if you're an elitist. And they tend to be conservative, in part because their sources of information—the Farm Bureau, Agritalk radio programs, and so on—are conservative."

But farmers will reach out when they find themselves up against the offal of a hog factory. An average hog produces two to four times as much raw sewage as a human being, so the 80,000 pigs raised by one of Premium Standard's operations in Lincoln Township, Missouri, for instance, can actually create as much waste as a city with 300,000 residents. In fact, it's estimated that the nation's 60 million hogs produce about 100 million tons of feces and urine each year.

Premium Standard is one of a handful of corporations—Seaboard, Tyson, and Smithfield are others—that so dominate hog production that many family farmers can only hang on by contracting with them to grow the companies' animals. Increasingly, though, communities are realizing that hog farms aren't worth the environmental damage, and are rejecting the overtures of the megafactories.

In Oklahoma, rancher and state senator Paul Muegge, persuaded by arguments from the Sierra Club and family farmers' groups, became an environmental hero when he authored the strongest set of regulations on hog production in the nation. In Alabama, Sierra Club members joined agriculturists in a protest calling for stronger regulations of mass-production animal farms. And in Kansas, farmers and conservationists in Great Bend managed to rebuff Seaboard's attempts to locate a factory there. Next the corporation tried St. Joseph, Missouri, where the city council acceded to community pressure and voted to keep them out. Then Seaboard sweet-talked politicians in Elwood, Kansas, but the community again thwarted the corporation. As Midkiff puts it, "Seaboard is still looking for a home, wandering the plains in vain."

Brian Blain and Richard Garcia's home is the nation's second-most-productive farming county. Farmers and environmentalists here in Tulare County, California, came together as a group dubbed POWER (Preserving Oaks, Water, Environmental Rights), and stared down the Tulare Irrigation District, an agency accustomed to moving water where and when it wanted.

For decades water was cheap for California farmers. But by 1998 the cost had risen to $34 per acre-foot. The Tulare Irrigation District figured it was losing more than $300,000 annually to seepage, so it decided to line its main intake canal with concrete.

The district had a problem, though: The canal, built to carry water from the Kaweah River watershed in the 1870s, was actually a series of natural channels connected by ditches. Many stretches retained riparian forests that would wither without groundwater; farmers and nearby towns that relied on the canal to replenish their aquifer worried that wells would dry up. The Kern-Kaweah Chapter of the Sierra Club opposed the project, and filed a lawsuit.

Meanwhile, Tulare County citizens, drawn by newspaper accounts or by word of mouth, were joining POWER, which included the Sierra Club. Even Bob Ludekens, owner of the 1,100-acre L. E. Cook Nursery and the Farm Bureau's Man of the Year, joined up. The previously unlikely allies, Ludekens, Blain, and Garcia, began slide and lecture presentations to service clubs, professional groups, city councils, chambers of commerce ("The lair of the enemy," grins Blain), and the county board of supervisors. In public confrontations with the water agency, POWER's team learned to shape its message to the audience. When talking to builders, the M-word ("moratorium"-if town wells dry up, building may be prohibited) was powerful; with conservationists, "habitat"; with workers, "jobs." "To create a diverse coalition, you have to show everybody that they are impacted," explains Blain. "Water was an issue that united us and nobody—conservative or liberal—wanted to see those trees removed."

Walnut-growers Don and Peggy Peterson found themselves swept into activism, and remain a little stunned by it. "We're grandparents," says Peggy. "Who would've thought we would be out there with picket signs."

Members of POWER raffled quilts, sponsored bake sales, and hosted barbecues to raise money for legal costs. "Little old ladies can have all the fundraisers they want, but we're still going to line this ditch," irrigation-district manager Gerald Hill reportedly said, words that would come back to haunt him when women started carrying signs reading "Little Old Lady Power." The day before a hearing on lining the canal was scheduled, the district foolishly announced it would begin bulldozing oaks along a section known as Potter's Slough. Richard Garcia quickly organized "Potter's Slough Blockade," a moving barricade that refused to allow district trucks and equipment to cross private property.

Equally important, the local press reported the encounter. "All of a sudden photos of prominent local citizens were on the front page and people said, 'That's Peggy. That's Bob. That's Sandy. That's my scoutmaster. That's my barber,'" Blain recalls. "'If they're involved, then there's really something to this.' Public sentiment kept swinging our way." Eventually, says Garcia, "we just wore them down. They had the money, but we had the people." In April 2001, the irrigation district shelved its plan to line the canal.

Sometimes the key to reaching agriculturists is understanding the financial challenges they face. "For farmers, the bottom line is trying to survive," explains organic farmer Chris Korrow of Burkesville, Kentucky. Korrow is an enthusiastic promoter of sustainable agriculture, which involves many specific practices—crop rotation for farmers and alternating grazing for ranchers, the avoidance of pesticides, and the reintroduction of native plants or animals, among others.

All seek to create agriculture that is economically viable and ecologically sound.

Korrow readily points out that only 8,000 of the country's farms are "certified organic." But that doesn't dampen his desire to bring sustainable agriculture to the rest. "We tell farmers a few simple things," Korrow says. "Sustainable farming is the only aspect of farming that's on the rise—organic sales are growing twenty-five percent a year—and that their inputs will be less and their profits higher." Even when sounding more like a banker than the eco-evangelist that he is, Korrow says it's still a hard sell. "These guys just want to make a living," he says. "And when extension agents tell them that organic is 'alternative,' to them that means 'risky.'"

There is one argument that hits home, however: "Being successful opens people's eyes," Korrow says.

But farmers don't always need an economic incentive to see the environmental light. On California's north coast, stream restoration and habitat protection are sources of great local pride. When wild salmon were designated endangered species in the mid-1990s, Sonoma County activist Kurt Erickson predicted, "The coho will come back when the community comes back together, because then the watershed will be healthy enough to sustain them." Indeed, citizens' groups have united a whole community—ranchers, farmers, winemakers, developers, and environmentalists—to protect and restore their nearby streams.

One Sonoma waterway has been reborn thanks to a visionary teacher named Tom Furrer. Petaluma, California, a suburb 40 miles north of San Francisco, was once a rural village veined by small but prolific streams. When in 1981 native-son Furrer returned to his hometown to teach at Casa Grande High School, he noted that the stream running next to the campus appeared moribund, with virtually no water, little foliage, and few fish. Adobe Creek had once hosted steelhead and salmon runs, but stream diversion and overgrazing, as well as the expansion of housing tracts, seemed to have pushed it beyond hope of revival.

Then Furrer encountered a rancher who had planted trees to shade his stretch of the stream and was trying to save a few steelhead fry in a pool there. That there were fry at all surprised Furrer, so he asked a simple question: Why not try to save the whole creek? He would end up educating a community about the health of streams and ecosystems, and about stewardship.

In 1983, Furrer's wildlife and forestry class began cleaning the creek and replanting the long-gone riparian forest. Despite setbacks (in 1989, for instance, county workers bulldozed 200 fledgling redwoods planted by the students), a circle of supporters arose within the community.

The public sector caught on late, but in 1992 the city of Petaluma joined the team by ceasing its diversions of the creek's water; for the first time in 80 years, Adobe Creek flowed naturally.

It took many car washes and candy sales and donations, but the group now known as the United Anglers of Casa Grande has annually planted 1,200 native willows, broad-leaved maples, and oaks; removed over 25 tons of garbage from the streambed; and constructed a state-of-the-art conservation fish hatchery. Only unprecedented cooperation could accomplish those things.

Just last year, for instance, cattle rancher Merv Sartori, through whose property Adobe Creek flows, agreed to allow more than four miles of wildlife-sensitive fencing to protect the stream from his 650 head of cattle. "The creek's going to overgrow because the cattle won't be there to beat it down," he says. "That's what the fish like, a lot of cover. It will keep the erosion down, no doubt about it."

Adobe Creek has come back to life because a community has come to life to support it. Shared problems can build teamwork, and common concerns can lead to cooperation. Once the doors are open even a crack, and folks begin to relate to one another as individuals rather than as stereotypes, they see that within familiar labels—farmer, rancher, environmentalist—there are plenty of variations, and plenty of ways to find common ground. ❖

Questions

1. Why is there mistrust between farmers and conservationists?
2. How much more raw sewage than a human being does the average hog produce?
3. How many of the country's farms are "certified organic"?

Answers are at the back of the book.

Introduced organisms are one of the largest threats to native species populations and have caused more than $130 billion a year in damage to the environment. The problem is a complex one; made more difficult by a lack of a comprehensive action plan to deal with introduced species. While the United States did create the National Invasive Species Council to deal with these issues, the council is continually thwarted by the fact that they must oversee multiple jurisdictions with multiple interests. The existing patchwork of state and local agencies nationwide lack a coordinating mechanism to prevent, eradicate, contain, and manage nonindigenous species in the U.S. Schmitz and Simberloff discuss a feasible model for such a coordinating agency.

Needed: A National Center for Biological Invasions

Don C. Schmitz, Daniel Simberloff

Issues in Science and Technology, Summer 2001

INTRODUCED ORGANISMS ARE THE SECOND GREATEST CAUSE, after habitat destruction, of species endangerment and extinction worldwide. In the United States, nonindigenous species do more than $130 billion a year in damage to agriculture, forests, rangelands, and fisheries, as estimated by Cornell University biologists. The invasions began in the 1620s with the inundation of New England and mid-Atlantic coastal communities by a wave of European rats, mice, insects, and aggressive weeds. Today, several thousand nonindigenous species are established in U.S. conservation areas, agricultural lands, and urban areas. And new potentially invasive species arrive every year. For example, the recently arrived West Nile virus now threatens North America's bird and human populations. In Texas, an exotic snail carries parasites that are spreading and infecting native fish populations. In the Gulf of Mexico, a rapidly growing Australian spotted jellyfish population is threatening commercially important species such as shrimp, menhaden, anchovies, and crabs. In south Florida, the government has conducted what the media calls a "chainsaw massacre, south Florida style": a $300-million effort to stop reintroduced citrus canker from spreading to central Florida by cutting thousands of citrus trees on private property.

A variety of local, state, and federal regulations and programs in the United States are aimed at restricting new invaders and managing and eradicating established ones. Unfortunately, however, the present response is highly ineffective, largely because it is fragmented and piecemeal. At least 20 federal agencies have rules and regulations governing the research, use, prevention, and control of nonindigenous species; several hundred state agencies have similar responsibilities. Within each state, hundreds of county, city, and regional agencies may also deal with nonindigenous species issues. A patchwork of federal, state, and local laws makes it difficult for these many agencies to manage existing invasions effectively and to prevent new ones.

During the past 20 years, government agencies and nonprofit organizations have attempted to solve coordination problems in the United States. However, these national coordinating interagency groups have been limited by their charters to specific regions or issues or have been understaffed or underfunded. Government agency and nonprofit staff working on these task forces or committees also have other responsibilities, so there are few working full-time on coordination. This lack of coordination and effectiveness as well as the dire nature of the threat necessitates a more powerful response: a new national center for biological invasions.

A Step in the Right Direction

Because of the growing economic and environmental impacts of biological invasions, President Clinton issued Invasive Species Executive Order 13112 on February 3, 1999, calling for the establishment of a national management plan and creating the National Invasive Species Council. The council, cochaired by the secretaries of Interior, Agriculture, and Commerce, includes the secretaries of Defense, State, Treasury, Transportation, and the administrator of the Environmental Protection Agency. An advisory committee recommends plans and actions to the council at local, state, regional, national, and ecosystem-based levels.

One of the National Invasive Species Council's major responsibilities has been the development of the National Management Plan on Invasive Species, released on January 18, 2001. The plan calls for additional funding and resources for all invasive species efforts and points out large discrepancies in funding across affected agencies. The plan also identifies problems in the current system, such as a failure to assign authorities to act in emergencies and the absence of a screening system for all intentionally introduced species. In addition, the plan calls for the National Invasive Species Council to provide national leadership and oversight on invasive species issues and to see that federal agency activities are coordinated, effective, work in partnership with the states, and provide public input and participation. The Executive Order specifically directs the council to promote action at local, state, tribal, and ecosystem levels; identify recommendations for international cooperation; facilitate a coordinated information network on invasive species; and develop guidance on invasive species for federal agencies to use in implementing the National Environmental Policy Act. Presently, the council has a staff of seven to accomplish these tasks.

The establishment of the National Invasive Species Council is an important initiative and reflects increasing U.S. investment in solving the problem of biological invasions. Although the council's management plan can be viewed as a federal coordination blueprint, there are some significant limitations on how the council will be able to implement the plan. Without the infrastructure, support, resources, and mechanisms to synchronize the thousands of prevention and management programs that now exist from coast to coast, the council is unlikely to be more effective at coordination than are other federal interagency groups. Under the plan, the same federal agencies mostly retain their responsibilities and their legislative mandates and will rely on existing interagency coordinating groups, state and local agencies, state invasive species committees and councils, regional organizations, and various nongovernmental organizations. In addition, the plan does not specify how federal agencies will work with state and local governments, especially in terms of detecting problem species early enough, so that every affected region can rapidly attempt to eradicate and/or contain a new invader to avoid or minimize long-term control efforts.

Indeed, the council's plan retains the overall federal agency structure without suggesting a mechanism to integrate the multiple programs that deal with biological invasions. It often delegates responsibility habitat by habitat, or in some cases, species by species, to various agencies that have traditionally managed or prevented the establishment of specific species. For example, the U.S. Department of Agriculture Animal and Plant Health Inspection Service (USDA-APHIS) responds to large, vocal groups that pressure Congress and the agency to conduct emergency operations or eradication efforts for invading species affecting a specific agricultural product. Citrus canker, gypsy moths, medflies, witchweed, and exotic animal and poultry diseases all have constituency-based programs. Many of these programs are effective in reducing the threat of these types of invasions. But federal agencies for the most part devote few resources to introduced nonindigenous species that lack an economically affected constituency. According to the General Accounting Office, federal obligations to address invasive species in FY 2000 totaled $631 million, but the USDA accounted for 88 percent of these expenditures.

This approach is inefficient because in many instances individual nonindigenous species are at worst minor nuisances by themselves but become major pests through their interaction with other introduced species. For example, large ornamental Ficus (fig) trees from Southeast Asia were introduced into Florida in the early 1900s without their pollinating wasps and remained sterile until the mid-1970s. Since then, pollinating wasps have been introduced by unknown means for at least three fig species, and these species have now become invasive in the public conservation lands of south Florida. More recently, the U.S. Centers for Disease Control and Prevention (CDC) has determined that the West Nile virus is most likely to have arrived in the United States in exotic frogs and to have been vectored by a recently introduced Asian mosquito. One of its major carriers is a nonnative bird, the house sparrow. These exotic species would fall under the purview of different agencies in the present structure. This is a situation in which existing policy and government structure have not responded to increased understanding of the dynamics of biological invasions.

Cooperation and coordination among agencies are essential to the success of nonindigenous species prevention and management efforts in the United States. However, government agencies are notoriously attached to their programs and prerogatives and may not participate in, or may even object to, initiatives by outsiders. Consider the case of the ruffe, a small perchlike fish native to southern Europe that has become the most abundant fish species in Duluth/Superior Harbor (Minnesota/Wisconsin) since its discovery there in 1986. Federal and state agencies developed a program to prevent its spread eastward from Duluth along the south shore of Lake Superior by annually treating several entering streams along the leading edge of the infestation with a lampricide. But at the last moment, members of state agencies decided not to support the plan, because they feared the lampricide could damage other fish species. Since then, observers have discovered the ruffe in the Firesteel River in the Upper Peninsula of Michigan, the easternmost record in Lake Superior. The ruffe is expected to have major effects on important fish species, such as the yellow perch. The ruffe could cause fishery damages that may total $100 million once it becomes established in the warmer, shallow waters of Lake Erie.

Another example involves a recently discovered Asian swamp eel population less than a mile from the Everglades National Park that threatens to undermine federal and state efforts to restore this unique ecosystem. These eels are voracious predators of native fish and invertebrates. The U.S. Fish and Wildlife Service, with assistance from the U.S. Geological Survey (USGS), is trying to implement a containment plan that involves removing aquatic vegetation, electrofishing infested canals, and trapping over an extended period. But the Florida Fish and Wildlife Conservation Commission says that the Asian swamp eel is now a permanent part of Florida's fish fauna and does not support the federal containment efforts. Solving such cooperation dilemmas is a key challenge to successful prevention, eradication, containment, and management of nonindigenous species in the United States.

Useful Models

This problem of multiple jurisdictional response has occurred before in disease prevention and management efforts and in fighting forest fires in the United States. The CDC in Atlanta and the National Interagency Fire Center in Boise, Idaho, are good models for a new national approach to the problem of invasive nonindigenous species. The CDC's Epidemic Intelligence Service (EIS) prevents new invaders, monitors existing outbreaks, implements prevention strategies, and has the responsibility for coordinating prevention and management efforts with foreign governments, numerous federal

agencies, at least 50 state agencies, and thousands of local governments and private organizations. The EIS was established in 1951 and is composed of physicians and scientists who serve two-year assignments. They are responsible for surveillance and response for all types of epidemics, including chronic disease and injuries. The EIS has played a key role in the global eradication of smallpox, discovered how the AIDS virus is transmitted, and determined the cause of Legionnaires' disease. Currently, 60 to 80 EIS staff members respond to requests for epidemiological assistance within the United States and throughout the world.

The National Interagency Fire Center shows how disparate agencies can work effectively together. The fire center's controlling body, the Multi-Agency Coordinating Group, which consists of five fire directors, has no controlling figure. The participating agencies—the Bureau of Land Management, the Forest Service, the National Park Service, the Bureau of Indian Affairs, and the Fish and Wildlife Service—have agreed to a rotating directorship so that all agencies have a chance at leadership. No one agency's agenda dominates the center's overall mission. By taking a macro view of forest fires, the center implements a national strategy of quickly attacking fires when they are small. In addition, the group facilitates the development of common practices, standards, and training among wildfire fighters. This effective strategy used in fighting our nation's forest fires is needed to combat the introduction and spread of harmful biological invasions in the United States.

The need for a coordinating mechanism between disparate agencies is so great that some federal research agencies are now establishing collaborative programs. The Smithsonian Environmental Research Center in Edgewater, Maryland, and the USGS Caribbean Science Center in Gainesville, Florida, will work together to collect, analyze, and disseminate information about aquatic species invasions in the United States. These types of collaborative programs must be expanded to include all affected federal and state agencies if we are going to lower the environmental and economic costs associated with biological invasions in the United States. Congress should create and pass legislation authorizing and providing funding for the National Invasive Species Council to oversee the establishment of a new kind of structure that will be similar to the CDC's EIS and the National Interagency Fire Center.

A New National Center

This new National Center for Biological Invasions could serve five functions. First, it could help coordinate the early detection of and rapid response to new invaders between federal, state, and local agencies and help determine factors that might influence their spread. Second, the center could enhance coordination of existing prevention and control efforts. By functioning as a neutral party, the center could broker cooperative agreements between agencies. Third, the center could enhance information exchange among scientists, government agencies, and private landowners. Fourth, the center could integrate university-based research to optimize management and prevention activities. Finally, the center could use diverse communication methods for wider and more effective delivery of public education about biological invasions.

Because most invasive species research is conducted in universities, the center should be strongly linked with a university or university system. Connecting the new center to a major university could also broaden contacts among all workers in the field of nonindigenous species. This approach would work better than current informal networks facilitated by Internet contact, which are often remarkably disparate. For instance, scientists working in weed management and those working on the ecology of nonindigenous plant species publish primarily in different journals, go to different meetings, and participate in different bulletin boards. Most pure science professional societies do not even have nonindigenous species interest groups. Unification of efforts would make research more efficient by fostering communication instead of isolation. Access to the resources of a major university could facilitate the construction and maintenance of a registry of all scientists working on nonindigenous species, with brief descriptions of their current projects and bibliographies of previous research.

Because of the academic association of the center, all agencies using its services could rely on the scientific integrity of its recommendations. In a university setting, the center would be less susceptible than government agencies to lobbying from constituencies such as agricultural industry groups, environmentalists, or other political organizations. By establishing scientific objectivity, the center could also influence these lobbyists and organizations. It would be especially important to build a relationship with the pet and ornamental plant industries, for which introduced species currently play a huge, profitable role. After all, the last presidential attempt to restrict the introduction of exotic species into U.S. ecosystems, President Carter's 1977 Executive Order 11987 on Exotic Organisms, was mainly ignored because it met with strong opposition from agriculture, the pet trade, and other interest groups. Center staff could function as neutral facilitators in organizing workshops and conferences to forge cooperative agreements for prevention, eradication, containment, or management efforts.

A National Center for Biological Invasions would also be able to help coordinate the surveillance necessary to identify new invasions. Surveillance serves several purposes: It is used to characterize existing invasion patterns, detect new ones, suggest areas of new research, evaluate prevention and control programs, and project future agricultural and resource management needs. National surveillance requires adequate infrastructure; a set of consistent methods; trained personnel within federal, state, and local agencies; and a network of taxonomists who can identify new invaders. USDA-APHIS has an extensive system in place to detect animal pests, pathogens, and parasites of livestock and cultivated crops. However, they are less successful at detecting invasive nonindigenous plants. Efforts by APHIS to detect nonindigenous plant or animal species that may affect nonagricultural areas are often hamstrung by a lack of adequate resources and the will to expand into an area where they lack a strong constituency. A national center could provide the necessary infrastructure for more effective surveillance and ensure that all biological invasions are adequately addressed.

Most states have developed networks of trained personnel within agriculture departments that provide extension services and communication pathways to entomologists, weed scientists, and animal control experts to prevent harmful invaders from diminishing agricultural output. But it is still possible for these networks to misidentify new invaders, as illustrated by the confusion surrounding the 1991 infestation in California by the sweet potato whitefly when a number of scientists believed it was a different species. There is no government-wide uniform procedure at the federal or at most state levels that identifies newly introduced organisms and tracks existing invasions; nor is there a consistent system of reporting them once they are found or of deciding on control efforts and evaluating control success. In addition, most states lack a network of trained personnel to address biological invasions in natural areas. Because of this lacuna, information concerning the identity and number and identities of biological invaders in the United States is incomplete.

Many control methods are species-specific, and improper species identification can lead to the failure of these management programs. In addition to the problem of inconsistent procedures, there is a shortage of trained taxonomists across the country. National, state, and university taxonomic collections in the United States provide reference material for identifying and comparing species by maintaining records of known species and their ranges. But rapid and accurate identification of newly introduced species is impeded by the fact that fewer biologists now specialize in taxonomy. People confronted with a new invader often do not know whom to call to identify it, because they do not have a list of experts and their areas of taxonomic specialty. In response to these problems, a new center could establish criteria for reporting on new invaders. Because the center would not be associated with any one agency, it could explore creating a consistent reporting approach. This task could be accomplished by organizing networks of scientists and using established monitoring programs. Wherever possible, the center could build on existing capacities and partnerships, such as the National Agricultural Pest Information System plant and animal data bases, the USGS Biological Resources Division, and nongovernmental databases, and forge strong links with local and state government agencies. A set of mapping standards, plus uniform methods for reporting new invasions and for assessing the extent of existing ones, could be developed and made available through the Internet. Synthesis would be a key role of the center.

In order for elected officials and decisionmakers to respond to a problem, someone must define its economic impact. Economic analyses of past harmful introductions are of uneven quality. Projecting future economic costs is more difficult because of uncertainty about biological outcomes. Scientific ignorance, long time lags between introduction and invasion, and changes in the natural world only confound the problem of good economic analysis. Potential effects also vary with the species and environments involved. Despite these limitations, economic analysis provides a useful benchmark to guide decisionmakers. The proposed center could establish models that would accurately define the economic impact of biological invasions in the United States. The center could work with economists to survey all state and federal agencies along with private landowners that deal with nonindigenous species. In addition, the center could survey affected businesses. Economic models could be used to analyze these data.

Perhaps the most important responsibility of this National Center for Biological Invasions would be the integration of prevention and management efforts at the local level. The national management plan relies heavily on federal initiatives; local and state agencies, which conduct most of the present management efforts, are almost an afterthought. In Florida, the Department of Environmental Protection established a statewide network of eleven regional working groups composed of federal, state, and local agency personnel and of nongovernmental organizations to manage upland invasive nonindigenous plants at the local level. These working groups have mapped distributions of invasive species, developed management plans, set regional control priorities, and removed unwanted species. Thousands of acres of invasive plant species have been eliminated, restoring native ecosystem functions. The center could help establish and strengthen local initiatives such as Florida's to prevent new invasions and manage existing ones.

The establishment of the National Invasive Species Council is a good first step in focusing policymakers' attention on this long, mostly silent war against biological invasions in the United States. However, the council currently lacks the infrastructure, support, resources, and mechanisms to synchronize the thousands of prevention, management, and research programs that now exist. The problem of biological invasions is largely soluble if infrastructure is established that responds to the multijurisdictional aspects of fighting biological invasions. The second step should be for Congress to create a national center, loosely modeled on the CDC's EIS and/or the National Interagency Fire Center, whose mission is to enhance existing programs and facilitate coordination and cooperation between local, state, and federal agencies. The establishment of a National Center for Biological Invasions would not guarantee that new invasions would not occur in the United States, but it would ensure that we are better prepared to respond to new invasions and to manage existing ones. ❖

Questions

1. What is the greatest cause for species endangerment and extinction?
2. How much damage do nonindigenous species do in the United States each year?
3. When did President Clinton issue Invasive Species Executive Order 13112?

Answers are at the back of the book.

The World Bank and International Monetary Fund have encouraged indebted countries towards privatization of their water supplies in hopes of luring international investors into their economies. Water was officially recognized as an "economic good" by our world leaders at the Rio Earth Summit in 1992. Water is now being recognized as a profitable commodity rather than a human right in some areas. With privatization schemes on the rise, failure to provide adequate utilities oversight could result in rate hikes, job cuts, environmental concerns, and poor service to rural or remote communities.

Privatizing Water

Curtis Runyan

World Watch, January/February 2003

IN APRIL 8, 2000, ROBINSON IRIARTE DE LA FUENTE, a U.S.-trained captain in the Bolivian army, lifted his rifle and fired into a barricade line of protestors. The previous evening, Bolivia's President Hugo Banzer had imposed martial law and called in the military to clear the streets in Cochabamba, the country's third largest city. Despite the crackdown, thousands of people—retired workers, middle-class students, street children, small-scale farmers—had stormed the central plaza to continue their months-long demonstration against the government's privatization of the city's water works.

The World Bank had threatened to withhold $600 million in debt relief if Bolivia did not privatize its water utilities. So in September 1999, Cochabamba signed over control of its aging, inadequate waterworks to Aguas del Tunari, an international consortium led by the U.S.-based corporation Bechtel. Water bills in the average household rose in one month by 35 percent. Meanwhile service generally remained sporadic, with water running less than four hours a day in many parts of the city. In response, the Coordinadora for the Defense of Water and Life, a coalition of labor organizers, environmentalists, and social activists, organized a general strike that shut down the city.

Iriarte's shot instantly killed 17-year-old student Victor Hugo Daza, enraging the protesters. Hundreds more were injured. With the violence growing, the executives of Aguas del Tunari fled the city. Government officials called an emergency meeting and rescinded the water contract, saying the company had abandoned its 40-year, $2.5 billion concession.

The protesters declared victory. The Bolivian government returned control of the water system to the utility, and has involved the Coordinadora. "The people-not the leaders-said no to privatization of water because it is a resource we cannot live without," said labor organizer and head of the Coordinadora, Oscar Olivera, at a speech a few days later in Washington, D.C. But Cochabamba's water troubles are far from over.

Privatization of state-run industries and utilities has long been a prescription of the World Bank and the International Monetary Fund (IMF). But privatization "shock therapy," intended to help indebted countries lure investment from international corporations, has only recently been applied to public water systems. When world leaders at the Rio Earth Summit in 1992 recognized water as an "economic good," they were acknowledging the failure of governments to provide clean drinking water to more than 1.1 billion people. This international commodification of water has opened the door for corporations to prospect for profits from a crucial resource that many consider to be a human right.

"Water is a public trust and it is the responsibility of governments to see to it that all of their people have access to a safe and adequate supply," said Sandra Postel, director of the Global Water Policy Project in Amherst, Massachusetts. "But they haven't done this, so how will the job get done? Can the private sector be a helpful partner?" The answers to these questions are certain to be the center of fierce debates in the future, as more and more governments are now turning to privatization as a means to attract muchneeded investment in aging and inadequate water systems. There are now 36 countries-all in Africa, Asia, or the Middle East-that are water stressed, meaning that they do not have enough fresh water to meet the industrial, municipal, and food production needs of their people. Seven more countries-including Ethiopia, Iran, and Nigeria—will join the ranks of the water stressed by 2015.

With more and more demand for water and fewer available supplies, the water industry estimates that potential returns in the global water market could run around $1 trillion. The untapped market is vast: private companies currently provide less than 10 percent of the water services worldwide. The two largest water corporations, Vivendi and Suez Lyonnaise des Eaux (both based in France), each draw annual revenues of roughly $10 billion from water and wastewater services. Together they provide water to more than 200 million people in more than 120 countries.

While countries with generally well-run public water utilities like the United States have been slow to privatize water systems, many developing countries-faced with rapid urbanization and deteriorating water systems-are looking to the corporate sector for assistance. The World Bank estimates that providing infrastructure to meet increased demands for water in the next decade will cost more than $60 billion a year. Privatization has offered some governments the chance to attract the private capital and expertise needed to build and expand expensive water systems, and help connect the millions who currently make do without piped water.

Many poor communities currently rely on questionable water sources and expensive, small-scale private suppliers. In Abidjan, Cote d'Ivoire, poor families pay 5 times the municipal rate; in Dhaka, Bangladesh, they pay 25 times; in Cairo, Egypt, 40 times. Privatization of the water utility in El Alto, Bolivia-where poor families currently pay more than 10 times as much for their water from tanker trucks than wealthier residents with piped water-is connecting poor communities to the municipal water system and reducing their water bills.

"In many countries water is a paternalistic sector-people expect water delivered at a low cost or for free," said Keegan Eisenstadt, a hydrologist with U.S.-based consulting firm Development Alternatives. "But as cities grow, this expectation makes it difficult for public utilities to raise enough money to maintain current systems, let alone build additional infrastructure." Managed carefully, privatization can help utilities improve their services by providing the capital and expertise needed to, for example, repair leaks in water mains, expand connections to unserved communities, or improve billing systems. But where governments have failed to provide adequate oversight, the results have been explosive.

Privatization schemes around the world have resulted in drastic rate increases, significant job cuts, fewer environmental safeguards, dropped conservation initiatives, and halted service to poor or remote communities. "Because the companies are motivated by profit and not public service, they have no incentive to supply the poor with water," says Maude Barlow, a leading opponent of water privatization with the advocacy group Council of Canadians, in her report Blue Gold.

In Tucuman Province in Argentina, a Vivendi subsidiary was forced to cancel its contract after large rate increases and poor water quality led to a general strike against paying water bills. In Ghana, where the average income is barely over $1 a day, massive protests broke out after the government nearly doubled water rates to prepare for privatization under pressure from the International Monetary Fund. In Atlanta, Georgia, the city council has threatened to cancel its contract with Suez after numerous cases of water contamination, and a decline in efficiency (the company has significantly cut staff to reduce its operating costs).

"The potential advantages of privatization are often greatest where governments have been weakest and failed to meet basic water needs," finds the Pacific Institute for Studies in Development, Environment, and Security in its thoughtful new report on water privatization. "Unfortunately, the worst risks of privatization are also where governments are weakest, where they are unable to provide the oversight and management functions necessary to protect public interests."

"Water privatization can foster corruption and result in rate hikes, inadequate customer service, and a loss of local control and accountability," said Wenonah Hauter of Public Citizen, a public advocacy group based in Washington, D.C. After years of lobbying from groups like Public Citizen, local unions, and other organizations, in November the city of New Orleans, Louisiana, turned down bids from Suez and Vivendi for a 20 year, billion dollar water contract.

"There is little doubt that the headlong rush to private markets has failed to address some of the most critical issues and concerns about water," said Peter Gleick, lead author of the Pacific Institute's report, "The New Economy of Water." Privatization may leave poor and under-served communities neglected, undermine conservation programs, or degrade ecosystems through efforts to tap new water sources. Still, the report concludes that private control of some elements of water provision-especially contracts with strong public oversight and the participation of affected parties-may help provide basic water needs for people and ecosystems, improve the efficiency and productivity of water systems, and facilitate equitable access to water for poor communities.

"How do we mobilize the capital needed to put in the pumps, pipes, and wells that are needed, and do it in a way that is fair, affordable, and environmentally sound?" asks Postel of the Global Water Policy Project. "Ideally, I'd rather the corporations stay out of it," she says. "They are motivated more by profits than public service, and they are accountable first and foremost to their shareholders, not the public. But that means governments need to step up to the plate."

Back in Cochabamba, the future of the city's water system is uncertain. While the Coordinadora's water cooperative has established new connections to a number of underserved communities and set up a huge new water tank, the system is still in disrepair and the utility is saddled by debts. New foreign investment looks unlikely, and 40 percent of the city remains without water service. And Bechtel is suing for breach of contract, demanding $25 million for its lost investment. But the Cochabamba water war has struck a chord around the world and raised serious questions about who will control water. ❖

Source: Worldwatch Institute, *World Watch,* Vol. 16, No. 1, Copyright 2003, www.worldwatch.org.

Questions

1. How much money in debt relief did the World Bank threaten to withhold if Bolivia did not privatize its water utilities?
2. By 2015, how many more countries will join the ranks of the water stressed, meaning that they do not have enough fresh water to meet the industrial, municipal, and food production needs of their people?
3. Water privatization schemes around the world have resulted in what?

Answers are at the back of the book.

Three major American environmental groups have pursued legal action against the Environmental Protection Agency for failing to curb global warming. Citing the government's own admissions that global warming is both environmentally injurious and at least partially accelerated by humans, the lawsuit claims that the government has continued to stall on taking action to reduce greenhouse gas emissions for years. This lawsuit has been fueled by the increasing frustrations of environmentalists with the Bush Administration's lack of action towards reducing the impact of global warming.

32

Groups Sue Government Agency Over Global Warming

Jim Lobe

Global Information Network, December 5, 2002

AMID GROWING ANGER AMONG ENVIRONMENTALISTS over the record and intentions of President George W. Bush, three major U.S. environmental groups said Thursday they are suing his Environmental Protection Agency (EPA) for failing to curb global warming.

The lawsuit by the Sierra Club, Greenpeace, and the International Center for Technology Assessment (CTA) charges the EPA with violating the 1977 Clear Air Act by failing to limit air pollution caused by automobiles that "may reasonably be anticipated to endanger public health or welfare".

Despite growing impacts of global warming on human health and the environment, the three groups charged, the EPA has steadfastly refused to control automobile emissions, which contribute to global warming.

"It's time for the Bush administration to get its head out of the sand," charged Joseph Mendelson, CTA's legal director. "The EPA stalling tactics are doing real damage in the fight against global warming."

The lawsuit marks the latest expression of rising frustration on the part of environmental activists over the administration's failure to act, despite a report by its own scientists last June that concluded that the burning of fossil fuels for industry and automobiles was contributing heavily to the climate change that will itself wreak havoc on natural ecosystems throughout the United States.

Environmentalists also fear future administration plans, particularly now that Republicans have gained control of both houses of Congress. Last year, much of the administration's energy plan, particularly its hopes of opening the Arctic National Wildlife Refuge (ANWR) to drilling by U.S. energy companies, was held up by the Democratic majority in the Senate.

But Republican control of Congress should make it much easier for Bush to relax existing environmental laws and regulations over the coming two years, at the behest of energy and automobile companies and electrical utilities that contributed heavily to his presidential campaign in 2000.

In the Senate, for example, the new chairmen dealing with energy and the environment both support drilling in ANWR and have among the upper chamber's worst voting records on environmental protection.

In his first move since the elections, Bush proposed a substantial loosening of federal regulations under the Clean Air Act two weeks ago to permit old coal-fired power plants to upgrade their facilities without requiring them to install new anti-pollution equipment, as they must now do.

While the administration insisted that the change would encourage investment that would eventually result in cleaner air, environmentalists blasted the proposals as a major step back in the fight against air pollution, and a number of leading Democrats called for EPA Administrator Christine Todd Whitman to resign her post in protest.

Whitman, a former governor of New Jersey, has long urged Bush to toughen regulations governing the Clean Air Act and even to sign the Kyoto Protocol, the international accord that requires industrialized countries to reduce their greenhouse gas emissions some seven percent below 1990 levels by 2012. The United States currently accounts for about 25 percent of the world's total greenhouse gas emissions.

But Whitman has been largely sidelined by the administration. She even avoided appearing personally to announce the power-plant proposals as she would normally do, issuing a statement through her spokesperson instead.

Thursday's lawsuit was motivated by the EPA's failure to respond to a formal petition submitted to it three years ago that demanded the regulation of global warming pollutants under the Clear Air Act.

The EPA subsequently received some 50,000 comments on the petition, the vast majority of which strongly agreed that global warming should be addressed under those provisions of the Clean Air Act that require it to regulate air pollution that may endanger public health or welfare.

Yet, 18 months after the public-comment period closed, the EPA has yet to offer a formal response to the petition, let alone to enact rules regulating greenhouse-gas emissions as requested by the petitioners.

According to the lawsuit, which cites the government's own studies about possible impacts of global warming on ecosystems and human health, climate change is responsible for unstable weather patterns, floods, droughts, and outbreaks of tropical diseases, including the West Nile virus that raged through much of the eastern United States last summer.

Scientists says warming, if left unchecked, will cause potentially catastrophic rises in sea level, the melting of the polar icecaps, and the loss of unique ecosystems around the world.

"Under the Bush administration, the EPA has found time to weaken or threaten many crucial environmental protections that Americans take for granted," according to David Bookbinder, an attorney with the Sierra Club. "But it can't find time to get serious about the most pressing environmental problem in the world's history."

The lawsuit coincides with the launch this week of the administration's first phase of its strategy to deal with climate change, a meeting of hundreds of scientists here to map out a research plan designed to better assess the problem and more accurately predict the effects of certain policy changes.

But environmentalists and many of the scientists taking part in the exercise have said enough is known about the threat posed by global warming to warrant a decision to cap, if not reduce, U.S. emissions immediately.

"The Bush administration is asking for five more years of studies while the world is warming and regular people will pay the price," said Gary Cook, climate coordinator for Greenpeace.

"We are asking the courts to intervene and order the EPA to enforce U.S. environmental laws and take action to address global warming. ❖

Questions

1. Who brought about the lawsuit and what does it charge the EPA with?

2. What held up much of the administration's energy plan last year?

3. What country currently accounts for 25% of the world's total greenhouse gas emissions?

Answers are at the back of the book.

Facing losses in their profits and reputations, U.S. automakers are being pressured by shareholders to cut their vehicles' greenhouse gas emissions. The U.S. has failed to sign on to the 1997 Kyoto Protocol, a move which shareholders fear could become a future liability in the global automotive market. Automakers may now have monetary incentive to stop lucrative but environmentally detrimental practices.

33

GM and Ford Pressed to Cut Greenhouse Gases

Jim Lobe

Global Information Network, December 12, 2002

A COALITION OF RELIGIOUS BODIES and other concerned investors is calling on the world's two biggest carmakers, General Motors (GM) and Ford, to take more aggressive steps to cut greenhouse gas emissions from their plants and products by 2012.

The group, led by the Interfaith Coalition for Corporate Responsibility (ICCR), has filed resolutions with both companies to be taken up at their annual shareholder meetings next spring. ICCR represents more than 275 faith-based institutional investors with about $120 billion in total investment assets.

A number of other major institutional investors, including California's largest public employee pension fund CALPERS—which has about $150 billion in assets—are expected to back the resolutions, according to Sister Patricia Daly, the head of ICCR and the Tri-State Coalition for Responsible Investment. "We'll be reaching out to CALPERS," she said.

Institutional investors' support for resolutions on global warming has shot up in the past several years from an average of only about 3 percent of companies' shares in 1999 to 18 percent in 2002. Such strong showings have historically been taken very seriously by corporate managers, particularly at a time when investors are worried about the stock market's performance.

"We believe that both General Motors and Ford face material and reputational risk," Daly said Wednesday, "in their current failure to address and reduce carbon dioxide emissions", which are believed by most scientists to be responsible for global warming.

"The high greenhouse gas intensity of U.S. vehicle manufacturers undermines the competitive positioning of U.S. automakers both here and abroad as the world, including their competitors, moves forward to address climate change. This is not only about what is good for the environment; it is about what is good for GM and Ford shareholders," she added.

Greenhouse gas emissions from trucks, vans and cars currently make up about 20 percent of total U.S. emissions, which contribute 25 percent of annual emissions worldwide.

While most of the industrialized world has signed on to the 1997 Kyoto Protocol, which requires wealthy countries to reduce their emissions to around 7 percent below 1990 levels, the administration of President George W. Bush has refused to do so, insisting that substantial reductions in greenhouse emissions would harm the U.S. economy.

At the same time, the administration has moved to ease existing laws and regulations designed to curb emissions and has aggressively sought out new sources of oil and gas in the United States, including environmentally sensitive areas in Alaska and elsewhere. It is also searching for fuel sources abroad, particularly in Russia, the former Soviet republics of Central Asia, and West Africa.

The result is that U.S. automobile companies, which would have been under pressure to sharply improve fuel efficiency had Washington ratified the Kyoto Protocol, lack government incentives to move in that direction.

As a result, the carmakers are falling further and further behind competitors, particularly Japanese manufacturers, in developing new, energy-efficient automobile technology, a trend which could well diminish the long-term value of the U.S. companies in an increasingly globalised world, say engineers and environmentalists.

"Shareholders have a right to expect GM and Ford to shed their environmental liability by selling cleaner-running vehicles—instead of being stuck in the mud with yesterday's technology while the Japanese step boldly ahead," said Kevin Knobloch, director of the Boston-based Union of Concerned Scientists (UCS).

"Using available and emerging improvements to conventional technologies, Ford and GM can build a fleet of vehicles that average 40 miles per gallon by 2012. Hybrid vehicles can boost that average to at least 55 miles per gallon by 2020 without sacrificing safety, comfort or utility for customers," he added.

The resolutions filed with each company ask them to report to shareholders by August 2003 on: estimated current annual greenhouse gas emissions from their plants and products; how they can significantly reduce such emissions from vehicles by 2012 and 2020; and an evaluation of what new public policies would enable the company to achieve those results.

This is not the first time that Ford and GM have faced challenges from the ICCR and other investors concerned about their policies on global warming.

In 1998, shareholder resolutions called on the two firms to withdraw from the Global Climate Coalition (GCC), a group of powerful U.S. oil, coal, gas, chemical and automobile companies that actively lobbied against Kyoto and tried to debunk claims that greenhouse emissions were contributing to global warming.

Although those resolutions received only 3 or 4 percent support from the two companies' shareholders, both giants decided to leave the GCC in 2000, leading an exodus of other firms that ultimately resulted in the GCC's collapse earlier this year.

"They were the key domino," noted Doug Cogan, deputy director of the social issues service of the Investor Responsibility Research Center (IRRC), an independent firm that tracks proxy voting.

Cogan noted that support for global-warming resolutions has mushroomed over the past three years and predicted that the Ford and GM resolutions will be approved by well over 10 percent of the shareholders.

The recent performance of global-warming resolutions, he said, "demonstrates much greater institutional backing of these proposals than in years past, and ranks them among the top vote-getting issues on social and environmental topics".

Last spring, support for a resolution that called for ExxonMobil to invest more in renewable energy drew more than 20 percent support, including from such heavyweight institutional investors as CALPERS.

Daly stressed that the new resolutions' supporters are eager to engage management in both companies about these issues, as they did in the late 1990s over their membership in the GCC. "This is going to take a real collaborative effort, and we're prepared to do what it takes," she stressed.

The Canadian parliament's ratification of the Kyoto Protocol on Monday is expected to bolster the insurgents' case, particularly because Canada is not only a major market for the two manufacturers, but also because it is home to many GM and Ford plants.

"The companies would do very well to take note of that vote," said UCS's Knobloch. ❖

Questions

1. The Interfaith Coalition for Corporate Responsibility has asked General Motors (GM) and Ford to cut greenhouse gas emissions by when?

2. What makes up about 20% of the United States' total greenhouse gas emissions?

3. What does the 1997 Kyoto Protocol entail?

Answers are at the back of the book.

Under the guise of economic and social process, free-trade agreements have had the effect of further elevating the status of transnational corporations to dangerous proportions. While producing cheap labor and inviting international consumer climates, free trade has also been producing economic and environmental threats to the earth's citizens. National environmental protections are being endangered for the sake of corporate property rights, prompting a vocal and growing opposition.

34

Tricks of Free Trade

Mark Weisbrot

Sierra, September/October 2001

FUTURE HISTORIANS WILL CERTAINLY MARVEL at how trade, originally a means to obtain what could not be produced locally, became an end in itself. In our age it has become a measure of economic and social progress more important even than the well-being of the people who produce or consume the traded goods. President George W. Bush recently declared free trade "a moral imperative." His predecessor, Bill Clinton, was prone to making wild economic claims for unfettered trade—for example, that it had added to employment and growth in the 1990s, contributing to the longest business-cycle expansion in American history. This is an economic and accounting impossibility, since our trade deficit, now running at a record $400 billion annually, actually ballooned during Clinton's presidency. Nevertheless, such assertions are rarely challenged in the press.

Technically, "free trade" refers to the absence of tariffs or other barriers that hinder the flow of goods and services across international boundaries. But it has recently morphed into a marketing tool to sell a whole range of new property rights for investors and corporations through an alphabet soup of sweeping international pacts: NAFTA, GATT, MAI, FTAA. In the last few years the environmental movement

has increasingly opposed these agreements. Together with organized labor, environmental groups were a major force in the collapse of the World Trade Organization's Millennium Round in Seattle at the end of 1999. More recently, they helped organize mass protests at the April 2001 "Summit of the Americas" in Quebec City.

Environmentalists were drawn into this debate because they were among the first to recognize that these trade deals were not primarily about "free trade" at all. For example, the most important provisions in the North American Free Trade Agreement (NAFTA) had nothing to do with the removal of tariffs, which were already quite low on goods imported from Mexico to the United States—about 2.5 percent on average. While Mexican tariffs on U.S. goods were higher, the Mexican economy was only one-twenty-fifth the size of ours. President Clinton did not spend months of his time, billions of tax-payer dollars (to win over NAFTA skeptics in Congress), and precious political capital fighting the rank and file of his own party just to open the relatively small Mexican consumer market to Big Macs and Krispy Kreme doughnuts.

The payoff for all this pork and political cliffhanging was not "free trade" but the exalted goal of a more secure investment climate for U.S. corporations. Under NAFTA, Mexico is bound by an international agreement that supersedes its own laws. Equally important, U.S. corporations got a safe haven of cheap labor where environmental regulations are rarely enforced.

In practice, however, NAFTA's biggest environmental threat turned out to be one that received little attention at the time the agreement was debated: Chapter 11, which allows foreign investors to sue governments directly for regulations that cause a loss of profits. This turned out to be a continental coup d'état for corporations, elevating them to the level of sovereign nations—something they had never achieved either under the General Agreement on Tariffs and Trade (GATT) or through the World Trade Organization (WTO). In the past, U.S. law has generally limited the definition of "expropriation" (for which the Constitution requires restitution) to government actions such as the taking of private land to build a highway. In the 1990s, the property-rights movement fought a (mostly unsuccessful) battle to broaden this definition to include what they called "regulatory takings"—for example, compensation for the reduced value of beachfront property due to environmental restrictions on its development.

But through NAFTA, in a solidaristic act of corporate internationalism, businesses and investors have granted each other what they couldn't win for themselves in their home countries. Chapter 11 allows companies that experience even a partial loss of profits because of regulatory action to seek reimbursement from the offending government. Consider the complaint brought under Chapter 11 against the state of California by Canada's Methanex Corporation over its gasoline additive, MTBE. Because MTBE is a known animal carcinogen, a possible human carcinogen, highly soluble in water, and very costly and difficult to clean up, it is seen as a major threat to groundwater. In California, more than 10,000 groundwater sites have already been contaminated by the additive. When California sought to ban MTBE, Methanex filed a Chapter 11 complaint. If the state wants to outlaw the substance, it may have to pay the company nearly a billion dollars.

A similar Chapter 11 case involving the Ethyl Corporation, the company that brought us the lead in leaded gasoline, turned the national tables. In 1997, the Canadian government banned the import of MMT, a manganese-based gasoline additive made by Ethyl that is a suspected neurotoxin, especially when its airborne particles are inhaled. "The history of leaded gasoline holds a very important lesson," says Elizabeth May, executive director of the Sierra Club of Canada. "If we want to put poisons in the blood and brains of our children, an excellent delivery mechanism is to add them to gasoline." Faced with a $250 million lawsuit brought by Ethyl, however, the Canadian government repealed its legislation banning MMT and paid the company $13 million in damages.

In 1995, the U.S. government led an attempt to extend this liberalized standard for takings—along with those three fateful words, "tantamount to expropriation"—to the 29 countries of the Organisation for Economic Co-operation and Development (OECD). The vehicle, a treaty known as the Multilateral Agreement on Investment (MAI), also sought to confer a host of other new rights and privileges on multinational corporations. But in doing so it sparked an enormous backlash against globalization, rallying more than 500 nongovernmental organizations against the proposed agreement. The treaty was almost complete before the American public became aware of its existence in 1996; within three years it was dead, largely because of this international campaign—one of the first, by the way, to be organized over the Internet. (See "All Hail the Multinationals!" July/August 1998.)

The body was dead, perhaps, but the soul migrated to the Free Trade Area of the Americas (FTAA). Proponents are portraying the new treaty as a helping hand to low-income countries because it creates a single open market that spans the hemisphere. But the helping hand is actually reaching out to corporations, offering them—as with the MAI—veto power over nations' environmental and public-health regulations.

The biggest threat posed by these commercial agreements and institutions is their usurpation of a nation's authority to rule in the interest of its own citizens. This is part of a long-term trend that has increasingly removed economic decision-making from parliamentary and other national institutions—which are at least potentially accountable to the wishes of an electorate—to unaccountable supranational bodies.

The most powerful of these by far are the International Monetary Fund and the World Bank. The IMF's clout comes from its position as the head of a cartel of creditors. (What OPEC is to oil the IMF is to credit.) A country that does not win the IMF's approval for its economic policy will be ineligible for most credit from the World Bank, other multilateral lenders, governments, and very often the private sector. While OPEC uses its control over oil resources to determine (as much as it can) the price of oil, the IMF uses its vast power to dictate economic priorities to dozens of developing countries.

The consequences are often disastrous for both the economy and the environment. For example, the IMF's and World Bank's advocacy of export-led growth, often based on nonrenewable resources, has caused enormous environmental destruction in countries that might otherwise have pursued more balanced growth strategies. Instead of developing local industries and talents, these countries are building gigantic dams, razing rainforests, and digging mines.

But it was NAFTA and the WTO that generated massive protests in the United States, where these institutions get most of their direction and are therefore most vulnerable. The huge popular rejection of the WTO came, in particular, because NAFTA failed so miserably to live up to the promises of its advocates. President Clinton and other NAFTA boosters claimed that the agreement would create new jobs in the United States, when instead it spurred hundreds of factories to close up and move south of the border. They also promised environmental improvement in Mexico, where in fact conditions have worsened. (See "Free-Trade Triage.")

Opposition was further galvanized by the new trade regimes' imperious insistence on their supranational authority. Should Europeans have the right to exclude hormone-injected beef from their markets? Most people would say yes, but the WTO ruled otherwise, even though the ban steered clear of protectionism by applying equally to both foreign and domestic beef. The scientific evidence did not justify the ban, the WTO ruled, thus substituting its own secretive deliberations for the judgment of the European Union's scientists and the desires of the European public.

More recently, another life-and-death issue has emerged to discredit the notion that "free trade" guides these institutions. A major objective of the WTO, NAFTA, and the proposed FTAA is to extend the enforcement of patents, copyrights, and other "intellectual property rights" beyond the borders of the wealthy countries where they are owned. A crucial test concerns the 36 million people who now have HIV/AIDS, most of them in the developing world. The "triple-therapy" drugs now widely used in the United States can keep people with HIV/AIDS alive and relatively healthy for many years, but at a cost of $12,000 per person annually, a prohibitive price for those in the developing world. Recently, the Indian generic-drug manufacturer Cipla offered to provide these drugs for as little as $350 per year. This would make treatment possible for millions of people, and millions more could be saved with relatively modest amounts of foreign aid from the high-income countries.

The United States, backing its major pharmaceutical companies, has fought to prevent such widespread distribution of generic versions of these and other life-saving medicines. For example, it went to the WTO to challenge Brazil's laws dealing with the manufacture and import of generic AIDS drugs—laws that form an important part of Brazil's remarkably successful AIDS-treatment program, which has already saved 100,000 lives and has cut the number of AIDS-related deaths there in half. (Stung by international criticism, the United States announced in June that it was dropping its challenge to Brazil.)

Extending patent rights to life-saving pharmaceuticals is the antithesis of free trade. It is, in fact, the most costly and deadly form of protectionism in the world today. By any standard economic analysis, a patent monopoly creates the same kind of economic distortion as a tariff. The major difference is that while tariffs rarely increase the price of goods by more than 25 percent, patent-protected prices can be 10 or 20 times the competitive price. The pharmaceutical companies maintain that their enormous profits are needed to fund necessary research and development. This is partly true, under present arrangements. But it merely strengthens the case for shifting R&D for essential medicines to the public and nonprofit sectors, which already account for about half of all U.S. biomedical research. The waste and inefficiency of using patent monopolies to fund this work is simply no longer affordable—especially in the face of AIDS, a pandemic more devastating than any since the bubonic plague killed a quarter of Europe's population in the 14th century.

The major multilateral economic institutions such as the World Bank, IMF, and WTO are not only unaccountable to any electorate, they have fundamental goals at odds with environmental protection. This is true in the extreme for the WTO, which was formed in large part to make sure that environmental and other policy goals of national governments did not "unnecessarily" impede international trade and investment flows. Its main actors—the top government officials and corporate CEOs of the G7 (the United States, Japan, Britain, France, Germany, Canada, and Italy) would rather ditch the whole project than watch it evolve into something that would allow trade sanctions to be used to advance such aims as environmental protection or labor rights.

The same is true for commercial agreements such as the FTAA, which is also very much corporate driven. (CEOs of corporations such as IBM and Coca-Cola, for example, are allowed to comment on drafts of the agreement before they are made available to the general public.) For these folks, such deals are gravy: They can do just fine with the status quo, and it would be irrational for them to accept anything that restricted the freedoms that they presently enjoy.

The IMF and the World Bank are another story: They have multiple goals and are many times more powerful than the WTO. Because they have the authority to impose a host of policies on borrowing countries, often under the threat of economic strangulation, these institutions cause more environmental destruction in a typical month than the WTO has brought about since its inception.

A major obstacle to reducing the damage caused by these bodies is their image as bulwarks against global economic chaos. Proponents depict the WTO as the protector of poor countries, because it allows trade to take place under a "rule-based system." Similarly, the IMF is seen as a lender of last resort, the global analog to an individual nation's central bank—rescuing countries in crisis just as the U.S. Federal Reserve System would bail out a private bank to prevent a financial breakdown from spreading.

This vision, however, presumes that the world really does have a "global economy" rather than a collection of national economies. Eighty percent of what is produced in the world (88 percent in the United States) is not traded internationally at all, and while it is true that most nations have evolved regulatory institutions like our Federal Reserve to resolve some of the problems inherent in a system of unregulated markets, the IMF does not play a similar role at the international level. Nor can we expect it to do so; in fact, it is much more of a world anti-government than a world government, promoting privatization of the public sector and deregulation of trade and investment flows (with the exception, of course, of intellectual-property rights, where "world government" seems to be the goal).

Most environmental policy—like the economic policy to which it is generally tied—will continue to be made at the national level. In addition to stopping the FTAA and WTO, then, we must reduce the power of the IMF and World Bank to impose environmentally unsound policies (such as export-led growth) and projects (like the disastrous World Bank–financed oil pipeline through the rainforest of Cameroon). This strategy of "harm reduction" means breaking up the creditors' cartel that these institutions control and weakening their grip on the policies of borrowing countries—which would include the IMF and the World Bank canceling the debts of poor nations. We cannot realistically expect to see environmentally sustainable economic strategies adopted in the developing world so long as these institutions hold sway.

In taking on NAFTA and the WTO, the environmental movement found itself in a powerful alliance with organized labor. Challenging the World Bank and IMF would yield many more allies throughout the world, like the hundreds of millions of small farmers in poor countries, whose markets the WTO seeks to flood with subsidized food from the highly mechanized farms of the United States and Europe. The IMF and World Bank squeeze more debt service from the poorest nations than these countries spend on health care or education. And the whole experiment in globalization has been an economic failure, even ignoring the environmental costs. In Latin America, for example, income per person has grown only 7 percent over the last 20 years, as these economies have opened up and followed the IMF's "structural adjustment" programs. In the previous two decades, per capita income increased by 75 percent, more than ten times as much.

In the United States, the marketing of "free trade" may have won over press and pundits, but it has failed to impress the general population, which has also suffered under globalization. The real median wage in the United States today is the same as it was 27 years ago. This means that the majority of the American labor force has been excluded from sharing in the gains from economic growth over the last quarter-century, an unprecedented event in our history. When asked to describe their views on trade in a Business Week/Harris poll last year, only 10 percent chose "free trader." Fifty percent chose "fair trader" (that is, a supporter of trade that pays a living wage to producers), and 37 percent chose "protectionist"—a word that is never used positively in the mainstream media. Although there were mixed feelings about globalization in general, most people chose "protecting the environment" and "preventing the loss of U.S. jobs" as major priorities for trade agreements—putting them very much at odds with our policymakers and trade officials.

This is not to say that there is no need for international institutions. On the contrary, agreements of the sort embodied in the Kyoto Protocol on global warming are essential. But the institutions and agreements promising "free trade" have a very different agenda. There is nothing "free" about creating new property rights for corporations while eroding national environmental protections. ❖

Reprinted with permission of the author from the September/October 2001 issue of *Sierra* magazine.

Questions

1. What is "free trade"?
2. How many people currently have HIV/AIDS and where are they predominately located?

3. Although there were mixed feelings about globalization in general, what did most people choose as major priorities for trade agreements?

Answers are at the back of the book.

Arguing that multinational corporations exploit workers, degrade the environment, and impede democracies, many people are demanding that corporations be held culpable for their actions in the global community. While "Anglo-Saxon shareholder capitalism" encourages companies to solely act in the interests of shareholders, "stakeholder capitalism" philosophy argues that corporations are beholden not only to their shareholders but also to their workers and communities. The clash between the two perspectives has grown into a global debate about corporate social responsibility.

35

Lots of It About—Corporate Social Responsibility

The Economist, December 14, 2002

ON DECEMBER 10TH, George Bush attended the opening ceremony of "Business Strengthening America" (BSA), an organisation whose aim is "to encourage civic engagement and volunteer service in corporate America". Also present were Don Evans, secretary of commerce, Steve Case, chairman of AOL Time Warner, and Jeffrey Swartz, boss of Timberland, a socially responsible shoe maker.

The high-profile occasion is the first fruit of a meeting held at the White House in June, when the president called for greater corporate involvement in the community. The collapse of Enron and the rash of subsequent scandals have made all of American business seem like an ogre who leaves entire communities high and dry when things go wrong. BSA wants to change this image by helping businesses to "do well by doing good".

This issue of corporate social responsibility (CSR)—how responsible companies should be to those other than their own shareholders—is arousing heated debate, and not just in America. In Europe and Asia, the battle is often said to be between Anglo-Saxon shareholder capitalism, which says that companies should pursue exclusively the interests of their shareholders, and stakeholder capitalism, which acknowledges that companies are also responsible to their

workers and local communities—often by having representatives from both on their boards.

The debate has also become entangled with that about globalisation. One of the main charges that the anti-globalisation brigade hurls at multinationals is that they behave irresponsibly all round the world. They exploit third-world workers, trash the environment and challenge democratically elected governments.

In all these tussles, the left demands that more rules be applied to companies, to make them more responsible. The right fires back that governments already subcontract far too much of their social policy to companies, using them as vehicles to limit working hours (in France), to promote racial harmony (in America) and to clean up the environment (just about everywhere on the planet).

This debate does not lack hot air; but it does lack a sense of history. How irresponsible have Anglo-Saxon companies been in the past?

Start in the East Indies

In practical terms, all companies—even Anglo-Saxon ones—have to secure what Leslie Hannah, an economic historian who is now dean of Ashridge Management College in England, calls "a franchise from society". Sometimes that franchise entails specific obligations: Mr Hannah points to ICI's altruistic work designing nuclear bombs for the British government during the second world war.

The franchise comes partly from consumers and other pressure groups in the private sector. As long ago as the 1790s, Elizabeth Heyrick led a consumer boycott, urging her fellow citizens in Leicester to stop buying "blood-stained" sugar from the West Indies. The East India Company was eventually forced to obtain its sugar from slaveless producers in Bengal. Nowadays, the influence of non-governmental organisations, such as Greenpeace, on corporate policy is far more pervasive.

However, the most enduring force on corporate behaviour has been the state. The mantra that, at some ideal moment in the misty past, Anglo-Saxon companies had nothing to do with the state is codswallop. Far from being alien to government, companies are a product of it: it was from the state that they derived the privilege of limited liability. Until the 19th century, monarchs generally gave companies a "charter" only if they were nominally pursuing the public good (which could mean anything from building canals in Lancashire to colonising India).

Indeed, it was the fact that companies were creatures of the state that turned Adam Smith against them. Most pioneers of the industrial revolution also regarded limited liability as a charter for rogues. The Victorians changed the perception, largely because a business arose that needed far more capital than mere partnerships could provide: the railways. By 1862, Britain had made it possible to set up a company, without parliamentary permission, for just about any purpose. The rest of the world's big countries—France, Germany, Japan and the United States—soon followed suit.

There was, in theory, a clear split among them, however. In Britain and America, the new joint-stock companies were freed from any obligation other than to obey the law and pursue profits. By contrast, in continental Europe (and later in Japan), companies were asked to pursue the interests of their various stakeholders, notably those of the state. But how different in practice were the Anglo-Saxons?

In Britain, most of the early companies—with the notable exception of the railways—had few dealings with the government. The gospel of free trade meant that there were no tariffs to go to London to lobby for. Corporate independence was even more noticeable in America. Businessmen created modern organisations that spanned the continent more quickly than the government did. In 1891 the Pennsylvania Railroad employed 110,000 people, three times the combined force of the country's army, navy and marines. Such monoliths could lay down their own rules.

But this freedom was never absolute. Unions sprang up to defend workers' rights, and "muckraking" journalists detailed corporate abuses. "I believe in corporations," said Teddy Roosevelt. "They are indispensable instruments of our modern civilisation; but I believe that they should be so supervised and so regulated that they shall act for the interests of the community as a whole." The supervision did not just take the form of antitrust legislation (which eventually broke up Standard Oil in 1911), but also of rules on health, safety, working hours and so on.

The Wall Street crash, a succession of corporate scandals and the Great Depression shifted public opinion on both sides of the Atlantic even more in favour of restricting the corporate franchise. America brought trucking, airlines and interstate gas and electric utilities under government control. In Britain, there was widespread horror at the inability of private coal companies to provide showers for their workers: they too were soon taken into public ownership.

This tightening continued into the post-war period. The British government nationalised the commanding heights of industry and, as late as 1967, John Kenneth Galbraith was arguing that America was run by an oligopolistic industrial state. Since the 1980s, however, the pendulum has swung the other way. Anglo-Saxon business has become more independent, for which it has to thank not only liberalising politicians such as Margaret Thatcher and Ronald Reagan, but also a new generation of more aggressive and numerous shareholders, who have demanded that companies be run in their interests.

This is the capitalisme sauvage that so upsets the French, in particular. But even today, Anglo-Saxon companies are more constrained than many critics suppose. The past two decades have seen a huge surge in social regulation, as various "stakeholders" have organised themselves into powerful pressure groups. In the late 1980s, for example, more than half of America's state legislatures adopted "other constituency" statutes that allow directors to consider the interests of all their stakeholders, not just shareholders. Connecticut even passed a law requiring them to do this.

Good Deeds Aplenty

There is one clear lesson from this history. Anglo-Saxon companies have often willingly taken on social obligations without the prompting of government. Mr Hannah cites the traditions of the Quaker families who founded so many of Britain's banks and confectionery firms: they had regular meetings where they were expected to justify to their peers the good that their businesses were doing.

Nor has corporate social responsibility been the preserve only of a few do-gooders inspired by religion. The notorious "robber barons" built much of America's educational and health infrastructure. Company towns, such as Pullman, were constructed, the argument being that well-housed, well-educated workers would be more productive than their feckless, slum-dwelling contemporaries.

Companies introduced pensions and health-care benefits long before governments told them to do so. Procter & Gamble pioneered disability and retirement pensions (in 1915), the eight-hour day (in 1918) and, most important of all, guaranteed work for at least 48 weeks a year (in the 1920s). Henry Ford became a cult figure by paying his workers $5 an hour—twice the market rate. Henry Heinz paid for education in citizenship for his employees, and Tom Watson's IBM gave its workers everything from subsidised education to country-club membership.

Critics tend to dismiss all this as window-dressing. But Richard Tedlow, an historian at Harvard Business School, argues that they confuse the habits of capital markets with those of companies. Capital markets may be ruthless in pursuing short-term results. Corporations, he says, have always tended to be more long-termist.

Alfred Chandler, the doyen of business historians, points out that the history of American capitalism has largely been the history of managerial rather than shareholder capitalism. For much of the 20th century, companies were run by professional managers; until the 1980s, institutional investors seldom made their presence felt. Today, more bosses are sacked, but managers still largely ignore shareholders. In Delaware, where around half of America's big companies are headquartered, the corporate code is more sympathetic to managers than it is to owners.

This has often made socially responsible behaviour easier. Rosabeth Moss Kanter, a professor at Harvard Business School, talks about most parts of the Anglo-Saxon world having a "CEO club" that successful businessmen are desperate to join. The price of admission is doing your bit for society.

But this cynical approach can be taken only so far. Most companies try to do good because they genuinely believe that taking care of their workers and others in society is in the long-term interests of their shareholders. The most successful Anglo-Saxon companies have consistently eschewed short-termism in favour of "building to last". For more than half a

century, Silicon Valley's pioneering company, Hewlett-Packard, has been arguing that profit is not the main point of its business.

There are two reasons why acting responsibly is in shareholders' interests.

- The first is that it builds trust, and trust gives companies the benefit of the doubt when dealing with customers, workers and even regulators. It allows them to weather storms, such as lay-offs or a product that does not work.
- The second is the edge it gives in attracting good employees and customers. Southwest Airlines is one of the most considerate employers in its industry: it was the only American airline not to lay people off after September 11th. Last year, the company received 120,000 applicants for 3,000 jobs, and it was the only sizeable American airline to make a profit.

Leave Well Alone

Since Anglo-Saxon companies have tended to be reasonably responsible in the past, without any government bullying, there must a good *a priori* case for leaving them alone. The case is strengthened by the fact that companies are most effective as social volunteers when they are doing things that are close to their shareholders' interests. Those interests clearly differ: oil companies tend to emphasise building local infrastructure; Avon, which sells products largely to women, is one of the world's biggest supporters of breast-cancer research. The idea of imposing one-size-fits-all laws is a mistake, for it undermines enthusiasm and prevents companies from exploiting their distinctive strengths. Why should all multinationals have to sign up to the same sort of mission statements that Shell and BP love so much?

In a sense, the corporate social responsibility movement has got it all the wrong way round. It tends to look for social goods—better housing, better schools, a cleaner environment—and then seek ways to force companies to provide them. It might do better to look at the things that the private sector does not deliver, and then get governments to fill the gaps. Left to themselves, companies will do more good than people give them credit for. But they are not here to build a fairer society. That is the job of government. ❖

Questions
1. By 1862, what had been made possible by Britain?
2. In 1891, how many people did the Pennsylvania Railroad employ and why is this significant?

3. How much did Henry Ford pay his workers?

Answers are at the back of the book.

36

Most economists, environmentalists, and other scientists perceive the relationship between the economy and environments to be one defined by finite limits. This paradigm states that there are biological and physical limits to economic growth, and if these limits are exceeded, ecological and economic collapse will be inevitable. Davidson instead proposes that the relationship between the environment and the economy be viewed as a continuum of degradation without clear limit points. He argues further that the concept of limits is harmful to environmental causes because there is an assumption inherent to the limits paradigm that degradation is quantifiable, while, in fact, degradation is too complex to be predicted. A full examination of these concepts follows.

Economic Growth and the Environment: Alternatives to the Limits Paradigm

Carlos Davidson

BioScience, May 2000

WHAT IS THE RELATIONSHIP between an expanding human economy and environmental quality? For most biologists, environmentalists, and ecological economists, the dominant paradigm for understanding the interactions between the economy and the environment is the concept of limits. The idea is that there are biological and physical limits to economic growth beyond which both ecological and economic collapse would occur. In this view, limits are seen as absolute constraints on economic activity, not just as a point beyond which economic growth results in environmental destruction. This concept of limits is a common theme, from limits on arable land (Malthus 1836), to energy and material limits (Meadows et al. 1972, 1992), to the economic scale and thermodynamic limits of ecological economists (Daly 1979, 1996). Although the limits concept has successfully been used to mobilize concern for environmental issues, the concept is problematic (Norgaard 1995). In this article, I argue that the concept of limits is ecologic ally and economically not useful and politically hinders the cause of conservation. I also propose metaphorical and analytical aspects of an alternative view.

Clearly, current human activities are causing environmental destruction at a scale and pace unprecedented in human history (Wilson 1988, 1992, Reid and Miller 1989). Moreover, any specific natural resource is finite and therefore there are absolute limits on its use. In addition, biological and physical systems underlie all economic activity and form constraints to which the human economy must adapt. However, I argue, contrary to the limits perspective, that biological or physical limits are seldom actually limiting to economic growth, such that reaching limits causes economic collapse or even stops growth. In most cases, the human economy is extremely adaptable and ways are found to adapt and continue to expand. Furthermore, in most cases, continued economic growth results not in ecological collapse but rather in continuous environmental degradation without clear limit points.

Whether or not environmental destruction is conceived of in terms of limits has important political implications. The limits perspective tends to focus on aggregate numbers of resources, consumption, and population and obscures the underlying causes of environmental destruction. I believe that examining the social structures of production and consumption offers greater hope for understanding and changing environmental destruction than does an analysis based on limits.

My arguments against the concept of limits build on and are in part similar to the arguments of Sagoff (1995). However, our approaches differ in key aspects. Sagoff, along with technological optimists (e.g., Simon 1981) and neoclassical economists (e.g., Nordhaus 1992), tends to discount the existence of environmental destruction and its negative impact on human welfare and to believe that new technology will allow the economy to expand without damaging the environment. My critique of limits, by contrast, is predicated on the assumptions that environmental destruction is real and that increases in the scale of the economy will contribute to greater environmental damage.

Traditionally, the term economic growth (or expanding economic scale) refers solely to the monetary value of output (i.e., gross domestic product, or GDP), which is not directly related to material use or waste production (e.g., $1 spent cutting timber, controlling pollution, and restoring a marsh all show up equally in GDP). However, I use the term to mean greater use of materials or increased waste production. I follow Daly (1996) in distinguishing economic growth (i.e., increased use of materials and waste) from economic development, which may provide for increased human welfare without increased use of materials or waste. I use the term environmental quality in its broadest sense to include biological diversity, resilience, and aesthetic, recreation, refuge, and ecosystem service values to humans.

Alternative Metaphors for Environmental Destruction

The relationship between economic activity and environmental quality is extremely complex. It is difficult to define, let alone meaningfully measure, the size of the economy or environmental quality. Consequently, our understanding of the interaction between the economy and the environment is primarily conceptual. Basic conceptual assumptions often take the form of metaphors (or conceptual models). Metaphors are not merely in the words we use but in the very concepts; therefore, they can shape the way we think and act (Lakoff and Johnson 1980). Often, we are not even aware of the content and power of our metaphors.

What is meant by ecological limits to economic growth can best be seen in the rivet metaphor developed by Paul and Anne Ehrlich (1981). In this well-known metaphor, an airplane is analogous to Earth. Each act of environmental destruction (loss of a species, in the original metaphor) is like pulling a rivet from the plane's wing. The wing has lots of rivets, so nothing happens when the first few rivets go. But eventually and inevitably, as more rivets are pulled, the wings break off and the plane crashes. In a related metaphor,

environmental destruction is likened to speeding toward a cliff in a car. If the car does not stop, it will eventually go over the cliff.

Three essential aspects of the rivet and cliff metaphors shape thinking about environmental problems. First, the transition from no effect to effect is abrupt. That initial changes have little effect contributes to a false sense of security and unwillingness to recognize limits and change course. Second, when limits are reached, the results are catastrophic—the plane crashes, the car goes over the cliff. Limits theorists generally predict that, if limits are reached or exceeded, there will be an ecological collapse which will in turn force a collapse of the human economy. Limits are seen as absolute constraints on economic activity, not just as points beyond which economic growth results in environmental degradation. For example, Ludwig (1996) writes, "Either we will limit growth in ways of our choosing or it will be limited in ways not of our choosing" (p. 16). The third essential component of these metaphors is that, in the event of a catastrophe, everyone suffers and therefore everyone has a clear self-interest in avoiding a crash.

The limits concept has been heavily criticized by neoclassical economists who believe that technical change will allow the economy to overcome all resource constraints and expand indefinitely (Nordhaus 1992). The basic neoclassical conceptual model, however, predicts either no environmental destruction or destruction only until the economy reaches a certain level of affluence; because of this prediction and others, this model has been criticized by ecological economists (e.g., Daly 1996).

A metaphor based on a tapestry provides a more accurate and useful view of the relationship between economic activity and the environment than either the limits metaphors of rivets and cliffs or the technological optimist model of neoclassical economics. Tapestries have long been used as metaphors for the richness and complexity of biological systems (e.g., the tapestry of life). As a metaphor for environmental degradation, each small act of destruction (akin to removing a rivet) is like pulling a thread from the tapestry. At first, the results are almost imperceptible. The function and beauty of the tapestry is slightly diminished with the removal of each thread. If too many threads are pulled—especially if they are pulled from the same area—the tapestry will begin to look worn and may tear locally. There is no way to know ahead of time whether pulling a thread will cause a tear or not. In the tapestry metaphor, as in the cliff and rivet metaphors, environmental damage can have unforeseen negative consequences; therefore, the metaphor argues for the use of the precautionary principle. The tapestry is not just an aesthetic object. Like the airplane wing in the rivet metaphor, the tapestry (i.e., biophysical systems) sustains human life.

However, the tapestry metaphor differs from the rivet and cliff metaphors in several important aspects. First, in most cases there are not limits. As threads are pulled from the tapestry, there is a continuum of degradation rather than any clear threshold. Each thread that is pulled slightly reduces the function and beauty of the tapestry. Second, impacts consist of multiple small losses and occasional larger rips (nonlinearities) rather than overall collapse. Catastrophes are not impossible, but they are rare and local (e.g., collapse of a fishery) rather than global. The function and beauty of the tapestry are diminished long before the possibility of a catastrophic rip. Third, there is always a choice about the desired condition of the world—anywhere along the continuum of degradation is feasible, from a world rich in biodiversity to a threadbare remnant with fewer species, fewer natural places, less beauty, and reduced ecosystem services. With the rivet and cliff metaphors, there are no choices: no sane person would choose to crash the plane or go over the cliff. This difference is key for the political implications of the metaphors. Finally, in the rivet or cliff metaphors, environmental destruction may be seen primarily as loss of utilitarian values (ecosystem services to humans). In the tapestry metaphor, environmental destruction is viewed as loss of utilitarian as well as aesthetic, option, and amenity considerations. (See Sagoff 1995 for a critique of conservation strategies that focus too narrowly on utilitarian values.)

Actual Environmental Destruction: Limits or Continuums?

How useful are the rivet and tapestry metaphors in describing actual experiences with the relationship between economic growth and environmental destruction? This question can be examined by looking at the variety of biological and physical limits to economic activity that have been proposed by ecologists, environmentalists, and ecological economists. In this article, I discuss five types of possible limits: input limits, limits on waste assimilation, entropy/thermodynamic limits, limits on human use of the products of photosynthesis, and limits attributable to the loss of biodiversity. The limits metaphor is a statement about the nature of both biophysical and human economic systems; therefore, limits need to be analyzed from both natural and social science perspectives. And, because human economies transport both inputs and wastes across the globe, the issue of biophysical limits to economic activity is best examined at a global scale.

Input limits. Until recently, input limitations received the most attention. Malthus (1836) predicted that limited

arable land would restrict the size of the human population through food shortages and starvation. Meadows et al.'s (1972) limits-to-growth models focused on a broader array of inputs but retained the basic Malthusian message: limited natural resources must limit human population and economic activity. Similarly, in The Population Bomb, Ehrlich (1968) predicted that hundreds of millions of people would starve to death in the 1970s from absolute food shortages. These predictions of absolute limits to the size of the economy due to resource exhaustion have repeatedly not been borne out. For example, despite over 150 years of predictions to the contrary, food production has consistently kept up with population growth. Between 1950 and 1985, total production of major food crops increased by more than 160%, more than matching population growth (Brown 1995). Millions of people starve or are malnourished every year, but not because of an absolute shortage of food (see Amartya Sen's [1981] classic Poverty and Famines).

Predictions of economic limits imposed by limited resources generally fail because they are based on the assumption that limits can be calculated according to current resource use and current resource stocks. This simple view of a limit is attractive but deceptive. Consider the example of cars, steel, and iron ore. A limit on the number of cars that may be produced cannot be calculated based solely on the amount of steel in a car and the size of known iron ore reserves. Car production depends on the amount of ore that is available from known reserves with current technologically and economically feasible extraction methods, the efficiency with which ore is converted to steel, the amount of steel required in a finished car, the efficiency with which the steel is used in producing cars, usage of steel for other products, and the rate of steel recycling. Any and all of these factors can and do change.

Production in capitalist economic systems is sufficiently flexible in substituting inputs that the scale of economic activity is not likely to be limited by input constraints any time soon. For example, in the 1970s, when energy prices in the United States increased dramatically, so did energy efficiency in manufacturing. Between 1973 and 1988, total energy use in US manufacturing declined by 13%, at the same time as output (value added) increased by 52% (Schipper and Meyers 1992). Even specific inputs do not appear to be as limiting as was once commonly thought. For example, between 1976 and 1996, proven reserves of crude oil increased by 65% and reserves of natural gas increased by 140% (OPEC 1997). Ultimately, the amount of any single input, such as oil, is limited, and even current levels of natural resource use have resulted in substantial environmental destruction (e.g., the collapse of fisheries and widespread deforestation). However, neither the fact that quantities of specific resources are limited nor the fact that resource use results in environmental destruction means that economic activity as a whole is limited by input constraints.

Waste absorption limits. In the 1980s, as the specter of aggregate material or energy shortages diminished, thinking on limits turned to the issue of waste absorption. Problems of waste absorption are potentially much more difficult to address than input constraints because pollution has the potential to cause irreversible and irreparable environmental harm and because there can be long time lags in detecting adverse affects. Furthermore, although economic incentives may at times encourage substitution for depleted inputs, economic incentives often also discourage reduction of pollution and encourage firms to locate in areas with lax environmental regulations (Daly 1996). For all of these reasons environmental degradation caused by waste production is a difficult ecological, technical, and social problem. However, the problem is not well illuminated by the concept of limits.

A limit for waste absorption analogous to the limits posited for inputs implies that only so much of a pollutant can be released in the environment before the environment will no longer absorb the waste, resulting in drastic negative consequences that ultimately curtail further dumping of wastes and that limit economic activity However, it is difficult to find documented cases that fit the waste absorption limit model. Although release of wastes often causes environmental destruction and may also have nonlinear effects (e.g., trophic cascades and algal blooms), catastrophic threshold points are seldom observed. Moreover, despite the protestations of industry, curtailing wastes often does not entail significantly limiting economic activity. If the environmental impacts of pollution tend to be gradual and continuous, then the concept of a limit for wastes has little meaning. Consistent with the tapestry metaphor, limit points and catastrophes are not ruled out, but they are probably rare. Carbon dioxide emissions and global climate change may be the best example of possible limit points for waste absorption. Although rising carbon dioxide levels may cause a continuum of impacts due to warmer temperatures and rising sea levels, there may also be catastrophic thresholds. For example, global warming could trigger large-scale changes in ocean circulation patterns that could in turn cause large and abrupt changes in climate (Broeker 1997). Past changes in ocean circulation patterns may have been responsible for the sudden ends of earlier interglacial periods.

This argument against the idea of waste limits is not that of the technological optimists, who deny that pollution is a serious problem. Clearly, pollution is causing massive environmental destruction and affecting human well-being. For example, widespread emissions of toxic chemicals may be responsible for soaring cancer rates. Industrial chemicals are found in the bodies of wildlife in even the most remote parts of the globe (Colborn et al. 1993). However, the fact that pollution is causing environmental degradation does not necessarily mean that there are catastrophic limit points. If there is a continuum of adverse effects, humans have to decide how much pollution we are willing to emit and what levels of environmental impacts we can live with. However, there may be no threshold point at which we must stop to avoid spiraling destruction.

Entropy and primary productivity limits. Herman Daly (1979, 1996) has developed a limits analysis that combines input and waste limits into constraints on throughput and the scale of the economy. Throughput is the total volume of material and energy flowing through the economy, starting as inputs and leaving as waste. Unlike Meadows et al. (1972) in The Limits to Growth, Daly does not assert that we are running out of material inputs. He recognizes the flexibility of production and does not want to tie limits to the use of any specific resource for which there may be substitutes. Instead, building on work by Georgescu-Roegen (1971), Daly appeals to limits on aggregate throughput based on thermodynamics and entropy, for which there is no substitution escape. The idea is that the earth and sun constitute a closed system. The total amount of matter and energy in the system is fixed and constant; however, there is a continuous, irreversible decline in the level of entropy. Humans use low-entropy energy from the sun and fossil fuel stocks and release high-entropy wastes. Early human societies relied primarily on energy from the sun; industrialized economies now depend primarily on the limited stock of fossil fuels.

Although entropy or thermodynamic limits are, theoretically, absolute, they are meaningful only if the human economy has a chance of approaching the limit. To be useful, the idea of entropy limits needs to be at least roughly quantifiable. What are the limits, and what is the size of the current global economy relative to those limits? Daly attempts to quantify these limits by referring to an analysis by Vitousek et al. (1986) of human use of net primary productivity (NPP). NPP is the solar energy captured by plants and other photosynthetic organisms minus that used by the organisms themselves for respiration. Vitousek et al. (1986) estimate that humans currently "appropriate" 25% of potential total global NPP and 40% of potential terrestrial NPP. Daly (1996) concluded that humans are therefore only 80 years away or less (two population doubling times) from appropriating the entire NPP, which he contends would be a biological disaster.

However, there are a number of serious problems with the NPP argument. First, human use of NPP is not an appropriate metric to assess possible entropy or thermodynamic limits. Entropy represents a theoretical limit to the economy because it encompasses all available energy. NPP, on the other hand, represents only a small fraction of even just the solar energy available on Earth. An entropy or thermodynamic limit to the economy implies that total human energy use is in danger of exceeding energy availability. Yet solar energy flow to Earth is many thousands of times greater than current global energy use (Dunn 1986). Although Daly (1996) appeals to entropy and thermodynamic limits, his NPP argument is more akin to earlier input limitation scenarios. The argument that NPP is an input limit suffers from the same flaws as other input limit arguments. Unlike entropy, total NPP is not fixed and may be increased in agriculture. More important, other inputs can be substituted for the products of primary producers: d irect solar energy can be used instead of firewood, and adobe, concrete, or steel can be used instead of wood for building materials.

In addition, Vitousek et al.'s (1986) estimates for human appropriation of NPP have been widely misconstrued as direct consumption figures and then used inappropriately to argue that NPP is an input limit to the economy (e.g., Goodland 1992). Appropriation of NPP in Vitousek et al.'s analysis is the sum of three separate categories: direct human use, co-opted NPP, and forgone NPP. Human consumption or direct use of NPP in the form of food, animal feed, timber, and fiber accounts for only 5.3% of appropriated NPP and only 1.4% of total NPP. Vitousek et al. (1986) measured NPP in petagrams (1 Pg = [10.sup.15] g) of organic material. Total global NPP was estimated to be 224.5 Pg and direct human use 7.2 Pg. The bulk (65.5%) of appropriated NPP comes from the co-opted category, which includes material "that is used in human-dominated ecosystems by communities of organisms different from those in corresponding natural ecosystems." (It should be noted that Vitousek et al. include direct use within the co-opted cat egory; for clarity I have maintained them as separate.) Thus, the entire NPP from the world's croplands (15 Pg) was counted as co-opted, even though direct use in terms of crops harvested is approximately only 1.8 Pg (Vitousek et al. 1986). Similarly, the entire NPP (9.8 Pg) from human-created pasture lands (e.g., human-created savannas in Africa and cleared pastures from forests in Latin America) is counted as co-opted, even though livestock consume only 0.7 Pg of NPP on these lands (Vitousek et al. 1986).

If NPP is envisioned as an input limit for the economy (Daly 1996, Ludwig 1996), then including co-opted NPP is inappropriate for assessing current human use. It is as if one counted all the water behind dams as co-opted, added the volume to that directly consumed, and, based on the total, asserted that there is a water shortage. Co-opted NPP is not consumed. Most of the NPP counted as co-opted flows to nonhuman organisms—albeit an altered set of organisms. Although it is clearly not desirable, there is no reason why humans cannot co-opt NPP production on all lands (resulting in 100% appropriation). Indeed, this may have already occurred, because to some degree humans have probably altered most of the planet (Vitousek et al. 1997).

The third component of NPP appropriation is forgone NPP, the loss of potential NPP due to land conversion. Vitousek et al. (1986) conclude that forgone NPP is 17.5 Pg, or 7.2% of total potential NPP (actual global NPP plus forgone NPP). However, roughly half of estimated forgone NPP results from the questionable assumption that the NPP of agricultural lands is less than that of the natural systems they replaced. Indeed, one of Vitousek et al.'s (1986) principal sources of global NPP data assumes just the opposite (Olson et al. 1983).

Although Vitousek et al. (1986) do not claim that products of NPP for human use are in danger of running out, they do suggest that human appropriation of NPP is leading to species extinctions because the vast majority of species must exist on the NPP that remains after human use. However, NPP appropriation probably does not provide a useful measure of human impact on the biosphere or threats to species' survival. Moreover, as an index of human impact on the environment (Vitousek et al. 1986) or of the size of the human economy (Arrow et al. 1995, Daly 1996), NPP appropriation may produce perverse results. For example, because NPP appropriation treats all human-altered lands as a loss, paving over a highly diverse traditional agricultural field does not show up as an increase in NPP appropriation. Instead, it only shifts the NPP of the agricultural field from the co-opted to the forgone category. In addition, increased carbon emissions and global warming may already be causing dramatic increases in NPP (King et al. 1997, Myneni et al. 1997), which would lead to a smaller percentage of NPP appropriated by humans and wrongly indicate a reduction in human environmental impact.

Biodiversity limits. The original rivet metaphor (Ehrlich and Ehrlich 1981) referred to species extinction and biodiversity loss as a limit to human population and the economy. A wave of species extinctions is occurring that is unprecedented in human history (Wilson 1988, 1992, Reid and Miller 1989). The decline of biodiversity represents irreplaceable and incalculable losses to future generations of humans.

Is biodiversity loss a case of limits, as suggested by the rivet metaphor, or is it a continuum of degradation with local tears, as suggested by the tapestry metaphor? In the rivet metaphor, it is not the loss of species by itself that is the proposed limit but rather some sort of ecosystem collapse that would be triggered by the species loss. But it is unclear that biodiversity loss will lead to ecosystem collapse. Research in this area is still in its infancy, and results from the limited experimental studies are mixed. Some studies show a positive relationship between diversity and some aspect of ecosystem function, such as the rate of nitrogen cycling (Kareiva 1996, Tilman et al. 1996). Others support the redundant species concept (Lawton and Brown 1993, Andren et al. 1995), which holds that above some low number, additional species are redundant in terms of ecosystem function. Still other studies support the idiosyncratic species model (Lawton 1994), in which loss of some species reduces some aspect of ecosystem function, whereas loss of others may increase that aspect of ecosystem function.

The relationship between biodiversity and ecosystem function is undoubtedly more complex than any simple metaphor. Nonetheless, I believe that the tapestry metaphor provides a more useful view of biodiversity loss than the rivet metaphor. A species extinction is like a thread pulled from the tapestry. With each thread lost, the tapestry gradually becomes threadbare. The loss of some species may lead to local tears. Although everything is linked to everything else, ecosystems are not delicately balanced, clocklike mechanisms in which the loss of a part leads to collapse. For example, I study California frogs, some of which are disappearing. Although it is possible that the disappearances signal some as yet unknown threat to humans (the miner's canary argument), the loss of the frogs themselves is unlikely to have major ecosystem effects. The situation is the same for most rare organisms, which make up the bulk of threatened and endangered species. For example, if the black toad (Bufo exsul) were to disappear from the few desert springs in which it lives, even careful study would be unlikely to reveal ecosystem changes. To argue that there are not limits is not to claim that biodiversity losses do not matter. Rather, in calling for a stop to the destruction, it is the losses themselves that count, not a putative cliff that humans will fall off of somewhere down the road.

The Politics of Limits

Is the limits metaphor a politically useful way to conceptualize environmental problems? If someone thinks that there is a cliff ahead in the road, she tells the driver, "There's a cliff." If that is not sufficient, she says, "It is a big cliff and we all are going to die if we go over." The limits approach assumes that "if only people understood" (i.e., saw the cliff and how big it is), they would stop their environmentally destructive practices (put on the brakes). After all, if the car crashes, everyone dies. All sane people are assumed to share a common interest in preventing a crash. The hope is that the existence and recognition of ecological limits external to society will force society to stop destructive practices. The limits perspective leads people to focus on pointing out limits and to emphasize the catastrophe that awaits if the limits are transgressed. As a consequence, writing about environmental degradation often has an apocalyptic tone.

Environmentalists have often predicted impending catastrophes (e.g., oil depletion, absolute food shortages and mass starvation, or biological collapse). This catastrophism is ultimately damaging to the cause of environmental protection. First, predictions of catastrophe, like the boy who cries wolf, at first motivate people's concern, but when the threat repeatedly turns out to be less severe than predicted, people ignore future warnings. Secondly, the belief in impending catastrophe has in the past led some environmentalists to support withholding food and medical aid to poor nations (Hardin 1972), forced sterilization (Ehrlich 1968), and other repressive measures. Not only are these positions repulsive from a social justice perspective, they also misdirect energy away from real solutions. And, by blaming poor and third world people for global environmental problems, these views have tended to limit support for environmentalism to the affluent in the first world. Fortunately, environmentalists of widely differing political perspectives, including some leading limits thinkers, now see alleviating human misery and poverty as essential to solving global environmental problems (Athanasiou 1996, Daily and Ehrlich 1996, Ehrlich 1997). In addition to recognizing the need to address poverty and inequality, recent limits writing has reduced its focus on catastrophe.

Historically, the limits metaphor has been part of a broader environmental and social analysis developed by authors such as Donella and Dennis Meadows, Paul and Anne Ehrlich, and Herman Daly. I refer to this broader analysis as the limits perspective. By focusing on aggregate quantities of natural resources, consumption, and population, the limits perspective depoliticizes our understanding of environmental destruction. What we consume, how much we consume, and how goods are produced are all political decisions that change over time and vary from country to country. Yet in the limits perspective, consumption and production technology are seen as more or less fixed, and significant social change is not even considered a possibility. In the most simplistic analyses, human population growth becomes the only variable in explaining environmental destruction. Similarly, many biologists who write on environmental issues erroneously apply the concept of carrying capacity to human society, and as a result ignore the social and political aspects of resource use. In animal populations, carrying capacity is the maximum population that can be sustained on the available resources in a given area. For human societies, however, carrying capacity has no real meaning unless consumption, technology, and a whole host of social variables are set at fixed levels (Cohen 1995). Viewing technology, consumption, and all social variables as fixed is implicit in the limits perspective, yet these variables are key to understanding the problem (Cohen 1995). For this reason, a recent high-profile statement of the limits perspective (Arrow et al. 1995) suggests moving away from the use of the carrying capacity concept.

The environmental destruction that is decried by the limits perspective is often real, even if it does not result from a transgressed limit, but there is something missing from this perspective. The focus on the cliff and catastrophe means that important political questions are often not asked: Why are we driving so fast? Who benefits from driving in this manner? Who has the right to decide how we drive and why? What views and beliefs support the current arrangements? Who benefits least from the current arrangements and might support change?

An Alternative Approach

The multiple threads of a tapestry together form a picture. Similarly, to better understand and challenge environmental destruction, it is necessary to examine the multiple factors shaping consumption and production and move beyond the singular focus of the limits perspective on aggregate population and resources. This approach means examining economic structures, social relationships of power and ownership, control of state institutions, and culture. For example, in the limits perspective, urban sprawl in western US cities is viewed as attributable principally or solely to population growth. Although population is an important factor, the limits perspective's focus on population leaves out other, equally important factors: economic incentives for developers to build large houses at low density, real estate interests' dominance of zoning and land-use planning decisions, and government funding for sprawl-inducing freeways instead of urban mass transit. All of these political, social, and economic factors are key for understanding sprawl, and, more important, for doing something about it.

The political-ecological approach is part of a growing body of research by geographers, anthropologists, economists, and biologists that draws on biological and social sciences to understand environmental problems. An excellent example is from Vandermeer and Perfecto (1995), who analyze the political and ecological causes and consequences of deforestation in Costa Rica. Other examples from very different perspectives include a collection by Painter and Durham, The Social Causes of Environmental Destruction in Latin America (1994), Richard Norgaard's Development Betrayed (1994) about the Amazon, and a recent critical review by Peet and Watts (1996).

Conclusions

The claim that, for the most part, there are not biophysical limits to economic growth may disturb many environmentalists. Dropping the limits/catastrophe paradigm is unattractive if one believes that appealing to people's rational desire to avoid a crash is the only way to motivate change and stop environmental destruction. The tapestry metaphor and the related political-ecological approach may be seen as pessimistic because they suggest that there are no external limits that are going to force a stop to environmental destruction. Without the threat of catastrophic limits, there is no guarantee of a fundamental commonality of interests to stop destructive practices. If environmental degradation is often gradual and continuous rather than catastrophic, then those in power who benefit materially from our current destructive economic system will fight to maintain the status quo.

However, the tapestry metaphor and the political-ecological approach have a hopeful side. Halting destructive processes is a political struggle that requires people to see beyond the aggregate numbers of resources, consumption, and population to understand the political, economic, and social forces responsible for environmental destruction. A political-ecological analysis often reveals that levels of consumption and destructive production processes are not fixed and inevitable but rather the result of political, economic, and cultural decisions that are subject to change. Environmental movements in many countries have been successful in bringing about significant changes, often against powerful political interests. For example, the US Clean Air and Clean Water Acts have greatly reduced air and water pollution. A political-ecological approach can illuminate possible solutions to environmental problems that may be obscured by the limits perspective. Finally, a political-ecological approach ties environmental issues to broader struggles for social justice and points to potential allies for conservation.

Acknowledgments

I wish to thank Paul Craig for encouraging me to write this paper and providing financial support to make it possible. Thanks to Craig, Cynthia Kaufman, and members of the Lorax political-ecology study group for helpful comments and numerous discussions that contributed to my thinking on limits. Comments by Richard Norgaard and two anonymous reviewers greatly improved the paper.

References Cited

Andren O, Clarholm M, Bengtsson J. 1995. Biodiversity and species redundancy among litter decomposers. Pages 141–151 in Collins HP, Robertson GP, Klug MJ, eds. The Significance and Regulation of Soil Biodiversity. Boston: Kluwer Academic Publishers.

Arrow K, et al. 1995. Economic growth, carrying capacity and the environment. Science 268: 520–521.

Athanasiou T. 1996. Divided Planet: The Ecology of Rich and Poor. Boston: Little, Brown.

Broeker WS. 1997. Thermohaline circulation, the Achilles heel of our climate system: Will man-made [CO_2] upset the current balance? Science 278: 1582–1588.

Brown LR. 1995. Nature's limits. Pages 3-20 in Brown LR, et al, eds. The State of the World. New York: Worldwatch Institute.

Cohen JE. 1995. How Many People Can the Earth Support? New York: W. W. Norton.

Colborn T, vom Saal FS, Soto AM. 1993. Developmental effects of endocrine-disrupting chemicals in wildlife and humans. Environmental Health Perspectives 101: 378–384.

Daily GC, Ehrlich PR. 1996. Socioeconomic equity, sustainabiity, and Earth's carrying capacity. Ecological Applications 6: 991–1001.

Daly HE. 1979. Entropy, growth and political economy of scarcity. Pages 67-94 in Smith VK, ed. Scarcity and Growth Reconsidered. Baltimore: Johns Hopkins University Press.

_____. 1996. Beyond Growth: The Economics of Sustainable Development. Boston: Beacon Press.

Dunn PD. 1986. Renewable Energies: Sources, Conversion, and Application. London: Peregrinus.

Ehrlich PR. 1968. The Population Bomb. New York: Ballantine Books.

_____. 1997. A World of Wounds: Ecologists and the Human Dilemma. Olderdorf (Germany): Ecology Institute.

Ehrlich PR, Ehrlich AH. 1981. Extinction: The Causes and Consequences of the Disappearance of Species. New York: Random House.

Georgescu-Roegen N. 1971. The Entropy Law and the Economic Process. Cambridge (MA): Harvard University Press.

Goodland R. 1992. The case that the world has reached limits: More precisely that the current throughput growth in the global economy cannot be sustained. Population and Environment 13: 167–182.

Grossman GM, Krueger AB. 1993. Environmental impacts of a North American free trade agreement. Pages 165–177 in Garber PM, ed. The U.S.-Mexico Free Trade Agreement. Cambridge (MA): MIT Press.

Hardin GJ. 1972. Exploring New Ethics for Survival: The Voyage of the Spaceship Beagle. New York Viking Press.

Kareiva P. 1996. Diversity and sustainability on the prairie. Nature 379: 673–674.

King AW, Post WM, Wullschleger SD. 1997. The potential response of terrestrial carbon storage to changes in climate and atmospheric [CO_2] Climate Change 35: 199–227.

Lakoff G, Johnson M. 1980. Metaphors We Live By. Chicago: University of Chicago Press.

Lawton JH. 1994. What do species do in ecosystems? Oikos 71: 367–374.

Lawton JH, Brown VK. 1993. Redundancy in ecosystems. Pages 255–270 in Schulze ED, Mooney HA, eds. Biodiversity and Ecosystem Function. Berlin: Springer-Verlag.

Ludwig D. 1996. The end of the beginning. Ecological Applications 6: 16–17.

Malthus TR. 1836. Principles of Political Economy. Reprint, Cambridge (UK): Cambridge University Press, 1989.

Meadows DH, Meadows DL, Randers J, Behrens WW. 1972. The Limits to Growth. New York: Signet.

Meadows DH, Meadows DL, Randers J. 1992. Beyond the Limits: Confronting Global Collapse, Envisioning a Sustainable Future. Post Mills (VT): Chelsea Green.

Myneni RB, Keeling CD, Trucker CJ, Asrar G, Nemani RR. 1997. Increased plant growth in the northern latitudes from 1981 to 1991. Nature 386: 698–702.

Nordhaus WD. 1992. Lethal model-2—The limits to growth revisited. Brookings Papers on Economic Activity 2: 1–43.

Norgaard RB. 1994. Development Betrayed: The End of Progress and a Coevolutionary Revisioning of the Future. London: Routledge.

_____. 1995. Metaphors we might survive by. Ecological Economics 15: 129–131.

Olson JS, Watts JA, Allison LJ. 1983. Carbon in live vegetation of world ecosystems. Oak Ridge (TN): Oak Ridge

National Laboratory, Environmental Science Division. Report no. ORNL-5862.

[OPEC] Organization of the Petroleum Exporting Countries. 1997. Annual Statistical Bulletin. Vienna (Austria): Organization of the Petroleum Exporting Countries.

Painter M, Durham W. eds. 1994. The Social Causes of Environmental Destruction in Latin America. Ann Arbor (MI): University of Michigan Press.

Peet R, Watts M. 1996. Liberation ecology: Development, sustainability, and environment in an age of market triumphalism. Pages 1–45 in Peet R, Watts M, eds. Liberation Ecology: Environment, Development, Social Movements. London: Routledge.

Reid WV, Miller KR. 1989. Keeping Options Alive: The Scientific Basis for Conserving Biodiversity. Washington (DC): World Resources Institute.

Sagoff M. 1995. Carrying capacity and ecological economics. BioScience 45: 610–620.

Schipper L, Meyers S. 1992. Energy Efficiency and Human Activity: Past Trends, Future Prospects. Cambridge (UK): Cambridge University Press.

Sen AK. 1981. Poverty and Famines: An Essay on Entitlement and Deprivation. New York: Oxford University Press.

Simon JL. 1981. The Ultimate Resource. Princeton (NJ): Princeton University Press.

Tilman D, Wedin D, Knops J. 1996. Productivity and sustainability influenced by biodiversity in grassland ecosystems. Nature 379: 350–363.

Vandermeer J, Perfecto I. 1995. Breakfast of Biodiversity: The Truth about Rain Forest Destruction. Oakland (CA): Institute for Food and Development Policy.

Vitousek PM, Ehrlich PR, Ehrlich AH, Matson PA. 1986. Human appropriation of the products of photosynthesis. BioScience 36: 368–373.

Vitousek PM, Money HA, Lubchenco J, Melillo JM. 1997. Human domination of Earth's ecosystems. Science 277: 494–499.

Wilson EO. 1988. The current state of biological diversity. Pages 3–18 in Wilson EO, Peters FM, eds. Biodiversity. Washington (DC): National Academy Press.

_____. 1992. The Diversity of Life. Cambridge (MA): Belkap Press of Harvard University Press. ❖

Questions

1. Between 1950 and 1985, by how much did the total production of major food crops increase?
2. Between 1973 and 1988, how did energy efficiency in manufacturing and energy prices in the United States change?

3. What is NPP?

Answers are at the back of the book.

Answers

SECTION ONE: THE ENVIRONMENT AND HUMANS

1. The Challenges We Face
1. The world leaders assembled to look at how to heal the ailing environment.
2. At present, 1.1 billion people lack access to clean drinking water and more than 2.4 billion lack adequate sanitation.
3. More than 11,000 species of animals and plants are known to be threatened with extinction.

2. Perceiving the Population Bomb
1. In 1990, the world was emitting, from the burning of fossil fuels and cement production, about 4.2 tons of carbon dioxide per person.
2. The maximum population that the Earth can accommodate, while allowing carbon dioxide emissions of 4.2 tons per person, is 2.1 billion people.
3. Both oil supply per capita and energy supply per capita peaked in 1979.

3. Rich vs. Poor
1. Between August 26 through September 4, 2002, the World Summit on Sustainable Development was held in Johannesburg, South Africa.
2. The Kyoto Treaty gets to the heart of consumption issues, because it encourages the development of clean societies and recognizes that the developed world is in the best position to do something to reverse the effects of climate change.
3. The focus on the free market worried many environmentalists because they fear that free-trade principles will supersede national environmental regulations.

4. Environmental Refugees
1. Flooding, drought, soil erosion, deforestation, earthquakes, nuclear accidents, and toxic spills have forced people worldwide to forever abandon their lands.
2. Dr. Norman Myers believes that climate change and environmental degradation will create 150 million environmental refugees by 2050.

3. Putting the brakes on climate change will only be achieved by reducing greenhouse emissions by 90 percent (*not* 10 or 20 percent) within a decade.

SECTION TWO: ENVIRONMENT OF LIFE ON EARTH

5. The Most Important Fish in the Sea
1. Menhaden make up approximately 40% of the catch of commercial fisheries.
2. Menhaden can be ground up, dried, and formed into another kind of kibble for land animals, a high-protein feed for chickens, pigs, and cattle.
3. Industry statistics show a dramatic decline in catches over the years since 1946.

6. Trout Are Wildlife, Too
1. Of the 14 named and unnamed cutthroat sub-species, two are already extinct, and the rest are in desperate trouble.
2. On Clear Creek, the cutthroat spawning run has declined from about 12,000 to 8,000 fish, and there is a corresponding decline at sample stations in the lake.
3. Antimycin is an incredibly selective and expensive fish poison that has a half-life of 40 minutes and is applied at 8 to 12 parts per billion.

7. Hostile Beauty
1. The Red Desert is an area of five million acres—bigger than the state of Connecticut.
2. Wyoming has a reputation as the nation's "energy breadbasket" because of its large reserves of coal, uranium, oil, and gas.
3. The Red Desert is a region that gets less than 10 inches of rain annually.

8. Wilding America
1. Wildlife corridors are ideal for a place like southern California because the prospect of creating large new preserves is relatively limited. Instead, many biologists believe it makes sense to simply connect smaller, established parks there.
2. The contemplation of wildlife corridors grew out of *The Theory of Island Biogeography*, written by Wilson and the ecologist Robert MacArthur in 1967.

3. In the chaparral, biologists are dripping the anal scent of bobcats onto rocks to lure felines toward camera traps.

SECTION THREE: RESOURCE USE AND MANAGEMENT

9. The Winds of Change
1. Experts say wind could provide up to 12% of the earth's electricity within two decades.
2. Some 1.6 billion people—a quarter of the globe's population—have no access to electricity or gasoline.
3. Iceland, which lies on a hotbed of underground volcanic activity, uses that geothermal energy to heat 90% of its buildings.

10. Scientists Say a Quest for Clean Energy Must Begin Now
1. Without prompt action, the atmosphere's concentration of greenhouse gases, mainly carbon dioxide from burning fossil fuels, is expected to double from pre-industrial levels by the end of this century.
2. Europe and Japan have accepted the Kyoto Protocol, a climate treaty, which includes binding deadlines for modest cuts in gas emissions.
3. The real solution, according to Kert Davies, is "cutting the use of fossil fuels by any means necessary."

11. Link Seen Between Water Scarcity and Poverty
1. About 10,000 government officials, representatives of international and non-governmental organizations, industry and water experts will attend the forum to discuss the world water crisis and its solutions.
2. The Water Poverty Index report grades 147 countries according to five measures—resources, access, capacity, use, and environmental impact.
3. Experts say that 20% of the world's population in 30 countries faced water shortages in the year 2000, a figure that will rise to 30% in 50 countries by 2025.

12. Atlanta's Growing Thirst Creates Water War
1. Georgia officials insist that they do not expect Atlanta to reach a real day of water reckoning until

2030, when they have projected that demands on the Chattahoochee will reach a maximum sustainable limit.

2. Unlike most American cities, Atlanta was founded, in 1837, not as a port but as a railway junction, which means that it is far from any major river or lake.

3. About 70% of the water supply for greater Atlanta is drawn from the Chattahoochee.

13. North America Losing Biodiversity, Say Experts

1. Tens of thousands of migratory birds are killed each year as a result of road building, cutting, bull-dozing, and burning by logging operations.

2. The total protected area in North America has increased by from less than 100 million hectares in 1980, to 300 million hectares, or about 15% of the continent's land surface.

3. For the most part, soil loss through erosion by wind and water has decreased because of better conservation practices and programs.

14. Buzz Cut

1. British Columbia agreed to prohibit logging on 1.5 million acres of the coastal Great Bear rainforest and to defer logging on an additional 2 million acres until a more environmentally sensitive forest-management plan is completed.

2. Seventy-five percent of British Columbia's cut comes from the east side of the Coast Mountains, much of it from the Chilcotin, and 90% of that goes to the United States, half as wood chips for paper and half as two-by-fours.

3. The U.S. Department of Labor has certified that Canadian imports have been a factor in the closing of more than 50 U.S. mills since 1996.

15. Feeding the World

1. In October 1999, the earth's population surpassed 6 billion people.

2. World food production increased 2.3% annually from 1961 to 1999, which outpaced the growth in population.

3. Africa's population will continue to increase, even under the United Nations' low/medium scenario, approaching 2 billion people in 150 years.

SECTION THREE: DEALING WITH ENVIRONMENTAL DEGRADATION

16. Ill Winds: The Chemical Plant Next Door

1. Polyvinyl chloride is used in tile, plastic water pipes, siding, and wire insulation.

2. The chemicals that one of Borden's biggest plants reportedly released into the air were ethylene dichloride, vinyl-chloride monomer, hydrogen chloride, hydrochloric acid, and ammonia.

3. Illiopolis' Borden plant can legally dump 800,000 gallons of wastewater into the stream every day.

17. Facing up to a Dirty Secret

1. Taiwan's Environmental Protection Agency found that the levels of toxic solvents in the water were up to 960 times higher than those considered safe for human use.

2. The RCA Taoyuan plant was in breach of safety rules governing the use of organic solvents because ventilation was insufficient, solvents were not labeled, advice on the effects of solvents on the human body was not posted, workers were not informed how to handle emergency situations, and routine and legally required health checks were not carried out.

3. California's Santa Clara County, the home of Silicon Valley, has 23 sites on the America's EPA National Priority List of the country's most polluted locations—more than any other county in the U.S.

18. Attacking an Arsenic Plague

1. Some 10 million private wells need testing in Bangladesh.

2. The arsenometer, an arsenic measuring device about the size of a Walkman, is rigged up from a pair of infrared LEDs, two glass tubes, a couple of photodiode detectors, an LCD display, and a 9-volt battery.

3. Fakhrul Islam, a chemist at Bangladesh's Rahshahi University, invented a cheap filter that strips arsenic from drinking water.

19. Long-Term Data Show Lingering Effects from Acid Rain

1. Progressively tougher pollution rules over the past three decades have reduced U.S. emissions of

sulfur dioxide by about 40% from its 1973 peak of 28.8 metric tons a year.

2. Some 15% of lakes in New England and 41% of lakes in New York's Adirondack Mountains are chronically or episodically acidic, and many such lakes have few or no fish.

3. If Congress were to call for an 80% SO_2 reduction from power plants below the current target for 2010, streams would probably bounce back by 2025, and some biological recovery in them might come by 2050.

20. News on the Environment Isn't Always Bad
1. 2050
2. Montreal Protocol
3. Causes of global warming arise from millions of individuals instead of a few corporations

21. The Weather Turns Wild
1. The newest global-warming forecast is backed by data from myriad satellites, weather balloons, ships at sea, weather stations, and by immense computer models of the global climate system.
2. In 1990, the cost for natural disasters was $608 billion, more than the four previous decades combined, according to Worldwatch Institute.
3. Fossil fuels remain far cheaper than the alternatives, but, to reduce this cost advantage, most Western European countries, including Sweden, Norway, the Netherlands, Austria, and Italy, have levied taxes on carbon emissions or fossil fuels.

22. Climate Policy Needs a New Approach
1. 75%
2. Poverty, poor land-use practices, degraded local environments, inadequate emergency preparedness
3. No

23. Bioreactors and EPA Proposal to Deregulate Landfills
1. Landfills are only liable for groundwater contamination 30 years after it is closed, just when the elaborate system of barriers are expected to fail, thus saddling future generations with the prospect of dealing with contaminated drinking water supplies at many locations.
2. If the EPA required incoming wastes to be treated and stabilized before burial, landfill tip fees would

be $65/ton, instead of the $20/ton at many of today's megafills.
3. "Bioreactors" is a technique of deliberately flooding the landfill with massive additions of liquids in an attempt to accelerate, rather than halt, decomposition.

24. Managing the Environmental Legacy of U.S. Nuclear-Weapons Production
1. The reactor accidents at Three Mile Island in 1979 and Chernobyl in 1986 raised public concerns about the continuing operations of U.S. production reactors.
2. Located in Nevada, Yucca Mountain has been recommended by the DOE as the site for a deep geologic repository for spent fuel and high-level waste.
3. Only a small portion of the $7 billion in annual funding is actually used for contaminant removal and waste processing. Most of the budget is spent on site surveillance and maintenance.

25. Silent Spring: A Sequel?
1. Scientists project average global temperatures will rise another 2.5 to 10.4 degrees F by the year 2100.
2. Global warming may also affect a bird's habitat because native forests are adapted to local climates, and many trees acclimated to cool environments are likely to shift northward.
3. Actions needed to slow global warming already are well known: reduce emissions of CO_2 and other greenhouse gases from fossil-fuel-burning power plants, factories, and automobiles.

SECTION FOUR: SOCIAL SOLUTIONS

26. Too Green for Their Own Good?
1. Some of the planet's most serious challenges include global warming, loss of biodiversity, and marine depletion.
2. The good news is that once an industry leader turns green, the rest often follow, fearful that consumers will punish them if they don't.
3. The World Conservation Union's most recent "Red List" indicates that about 24% of mammals "are currently regarded as globally threatened."

27. Seeing Green: Knowing and Saving the Environment on Film

1. *On Nature's Terms* argues that wolves, pumas, and bears are "keystone species," or important indicators of healthy ecosystems, because they require so much open space and regulate other animal populations. It argues that prejudicial attitudes, government policies, and suburban sprawl have all contributed to their extermination, but it presents how citizen action is conserving lands and contributing to species preservation. It provides no detail on the contours of Americans' changing attitudes toward predators and predator loss, much less the diverse political and social meanings people make of arguments about predators and their roles in ecosystems.

2. The films inevitably raise basic questions about what strategies will work to "save the environment," the forms of knowledge on which environmental practices are based and the social orders and hierarchies such practices imagine, and the role of the individual in promoting structural change.

3. Ecotourism serves as a strategy for sustainable economic development, cultural preservation, and nature conservation.

28. A Forest Path Out of Poverty

1. The Callari Project, a marketing cooperative spanning 15 towns and villages in Ecuador's Napo province, aims to help indigenous people make a living without destroying their forest or getting involved in the Colombian conflict.

2. Since the Callari Project formed, 700 artisans and 300 farmers have increased their incomes 30% by improving the quality of the cocoa and coffee they grow, relearning indigenous methods for producing useful items from jungle materials, and marketing their products abroad.

3. The cooperative was initiated by a Kansan named Judy Logback, who came to Ecuador to do environmental education in 1997 and tried to persuade indigenous people to stop cutting down their trees for sale.

29. Growers and Greens Unite

1. Conservationists are often lumped in with outsiders who don't understand what's really happening on the farm and often discount farmers as stubborn and narrow-minded.

2. An average hog produces two to four times more raw sewage than a human being.

3. Only 8,000 of the country's farms are "certified organic."

30. Needed: A National Center for Biological Invasions

1. Habitat destruction is the greatest cause, followed by introduced organisms, of species endangerment and extinction worldwide.

2. In the United States, nonindigenous species do more than $130 billion in damage per year to agriculture, forests, rangelands, and fisheries, as estimated by Cornell University biologists.

3. President Clinton issued Invasive Species Executive Order 13112 on February 3, 1999, calling for the establishment of a national management plan and creating the National Invasive Species Council.

31. Privatizing Water

1. The World Bank had threatened to withhold $600 million in debt relief if Bolivia did not privatize its water utilities.

2. Seven more countries, including Ethiopia, Iran, and Nigeria, will join the ranks of the water stressed by 2015.

3. Water privatization schemes around the world have resulted in drastic rate increases, significant job cuts, fewer environmental safeguards, dropped conservation initiatives, and halted service to poor or remote communities.

32. Groups Sue Government Agency over Global Warming

1. The lawsuit brought by the Sierra Club, Greenpeace, and the International Center for Technology Assessment charges the EPA with violating the 1977 Clear Air Act by failing to limit air pollution caused by automobiles that "may reasonably be anticipated to endanger public health or welfare."

2. Last year, much of the administration's energy plan, particularly it's hopes of opening the Arctic National Wildlife Refuge to drilling by U.S. energy companies, was held up by the Democratic majority in the Senate.

3. The United States currently produces 25% of the world's total greenhouse gas emissions.

33. GM and Ford Pressed to Cut Greenhouse Gases

1. The Interfaith Coalition for Corporate Responsibility asked General Motors (GM) and Ford to take more aggressive steps to cut greenhouse gas emissions from their plants and products by 2012.
2. Greenhouse gas emissions from trucks, vans, and cars currently make up about 20% of total U.S. emissions.
3. The 1997 Kyoto Protocol requires wealthy countries to reduce their greenhouse gas emissions to around 7% below 1990 levels.

34. Tricks of Free Trade

1. "Free trade" refers to the absence of tariffs or other barriers that hinder the flow of goods and services across international boundaries.
2. Thirty-six million people now have HIV/AIDS, and most of them are located in the developing world.
3. Although there were mixed feelings about globalization in general, most people chose "protecting the environment" and "preventing the loss of U.S. jobs" as major priorities for trade agreements, putting them very much at odds with our policymakers and trade officials.

35. Lots of It About—Corporate Social Responsibility

1. By 1862, Britain had made it possible to set up a company, without parliamentary permission, for just about any purpose.
2. In 1891, the Pennsylvania Railroad employed 110,000 people, three times the combined force of the country's army, navy, and marines, which meant they could lay down their own rules.
3. Henry Ford became a cult figure by paying his workers $5 per hour—twice the market rate.

36. Economic Growth and the Environment: Alternatives to the Limits Paradigm

1. Between 1950 and 1985, total production of major food crops increased by more than 160%, more than matching population growth.
2. Between 1973 and 1988, total energy use in U.S. manufacturing declined by 13%; at the same time output (value added) increased by 52%.
3. Net primary productivity (NPP) is the solar energy captured by plants and other photosynthetic organisms minus that used by the organisms themselves for respiration.